Sustainable Civil Infrastructures

Sustainable Civil Infrastructures (SUCI) is a series of peer-reviewed books and proceedings based on the best studies on emerging research from all fields related to sustainable infrastructures and aiming at improving our well-being and day-to-day lives. The infrastructures we are building today will shape our lives tomorrow. The complex and diverse nature of the impacts due to weather extremes on transportation and civil infrastructures can be seen in our roadways, bridges, and buildings. Extreme summer temperatures, droughts, flash floods, and rising numbers of freeze-thaw cycles pose challenges for civil infrastructure and can endanger public safety. We constantly hear how civil infrastructures need constant attention, preservation, and upgrading. Such improvements and developments would obviously benefit from our desired book series that provide sustainable engineering materials and designs. The economic impact is huge and much research has been conducted worldwide. The future holds many opportunities, not only for researchers in a given country, but also for the worldwide field engineers who apply and implement these technologies. We believe that no approach can succeed if it does not unite the efforts of various engineering disciplines from all over the world under one umbrella to offer a beacon of modern solutions to the global infrastructure. Experts from the various engineering disciplines around the globe will participate in this series, including: Geotechnical, Geological, Geoscience, Petroleum, Structural, Transportation, Bridge, Infrastructure, Energy, Architectural, Chemical and Materials, and other related Engineering disciplines.

SUCI series is now indexed in SCOPUS and EI Compendex.

More information about this series at http://www.springer.com/series/15140

Behzad Fatahi · Shen-En Chen ·
Jun Hu
Editors

Resilient Design and Construction of Geostructures Against Natural Hazards

Proceedings of the 6th GeoChina International Conference on Civil & Transportation Infrastructures: From Engineering to Smart & Green Life Cycle Solutions – Nanchang, China, 2021

Springer

Editors
Behzad Fatahi
Civil and Environmental Engineering
University of Technology Sydney
Sydney, NSW, Australia

Shen-En Chen
Civil and Environmental Engineering
University of North Carolina at Charlott
Charlotte, NC, USA

Jun Hu
School of Civil Engineering and Arch.
Hainan University
Haikou City, China

ISSN 2366-3405 ISSN 2366-3413 (electronic)
Sustainable Civil Infrastructures
ISBN 978-3-030-79853-6 ISBN 978-3-030-79854-3 (eBook)
https://doi.org/10.1007/978-3-030-79854-3

This Springer imprint is published by the registered company Springer Nature Switzerland AG
The registered company address is: Gewerbestrasse 11, 6330 Cham, Switzerland

Introduction

The rapid population growth accompanied by the urbanisation has led to the expansion of construction projects over marginal soils worldwide. Thus, it is crucial to employ new solutions for many construction activities to avoid the potential failure of the foundations and excessive post-construction settlements. This volume contains 12 papers accepted and presented at the GeoChina 2021 International Conference on Civil & Transportation Infrastructures: From Engineering to Smart & Green Life Cycle Solutions, held in NanChang, China, during September 18 to 19, 2021. It contains research data, discussions and conclusions focusing on numerical modelling and experimental studies in geotechnical engineering. Topics include issues related to civil infrastructure such as ground improvement techniques, seismic response and issue related to problematic and marginal soils. Research advances related to the above-mentioned topics help to quantify the uncertainties related to response of infrastructure to the applied loading conditions and help to optimise the design and construction. The novel solutions proposed for ground improvement presented in this book help to achieve the sustainability goals and offer green life cycle solutions in construction projects. Combination of research papers presented in this book covering issues from large-scale field studies to small-scale laboratory experiments supported by numerical modelling allows practising engineers and researchers access to state-of-the-art solutions helping to minimise the maintenance cost of infrastructure.

Contents

About the Editors

Prof. Behzad Fatahi is Head of Discipline (Geotechnical and Transportation Engineering) and Associate Professor in the School of Civil and Environmental Engineering. He has published more than 100 peer-reviewed journal papers related to civil/geotechnical/structural engineering. He has about 10 years of industry experience in the field of civil/ geotechnical engineering in consulting companies such as Coffey and Soil Engineering Services (SES) resulting in more than 50 major technical reports for clients. Behzad was named Australasia Young Railway Engineer of Year by Engineers Australia and Railway Technical Society of Australasia. Behzad was also awarded the first prize Young Geotechnical Professional by the Australian Geomechanics Society (Sydney Office). His research and practical interests include ground improvement techniques for roads and railway lines, viscous behaviour of clays, bioengineering, soil–structure interaction and unsaturated soil mechanics. Prof Fatahi has worked significantly on design of bridges, structures and foundations against earthquakes and has been researching on new design and construction techniques to improve performance of these structures under large earthquakes and fault ruptures.

Prof. Shen-En Chen is Professor of the Department of Civil and Environmental Engineering at the University of North Carolina at Charlotte. He is Registered Professional Engineer with a research interest in Structural Health Monitoring of Bridges, Power Transmission Structure Design, Geophysics and Mining Science. He completed both his undergraduate and graduate studies at the West Virginia University. He has authored, co-authored and edited 92 journal papers, 15 books and chapters and 80 peer-reviewed conference papers. He is Fulbright Scholar, Indian GIAN and TEQIP visiting professor (National Institute of Technology Trichy), high-end experts for Tongji University and Sichuan University of China.

Prof. Jun Hu is Associate Professor (MEng/PhD supervisor) of the School of Civil Engineering and Architecture of Hainan University and Director of Institute of Marine Geotechnical Research. He was appointed as the first batch of youth "Master of South China Sea" and key member of the "Double Hundred" talent person team in Hainan Province. He was also honoured as the top-notch person of Hainan Province. With over 130 publications in international refereed journals, more than 100 national patents, seven software copyrights, four monographs and four textbooks, he is currently conducting interdisciplinary research relating marine geotechnical engineering, tunnel and underground engineering, artificial ground freezing technology and intelligent construction.

Laboratory Evaluation and Neural Network Modeling of Treated Macau Marine Clay

Yuanhang Wang(✉), Wayne Yan, Jacqueline Ieong, and Thomas M. H. Lok

Department of Civil and Environmental Engineering, University of Macau, Macau SAR, China
mb75464@um.edu.mo

Abstract. The marine clay is generally regarded as poor materials for urban applications because of its high settlement and instability. In this study, the soft marine clay is mixing with cementitious material such as cement and lime in different ratios, so that the geotechnical characteristics of blends improved. A series of laboratory experiments are conducted to verify the enhanced performance of treated marine clay. The optimum moisture content (OMC) along with maximum dry unit weight obtained from Standard Proctor compaction test and Harvard miniature test. Unconsolidated undrained test (UUT) and unconfined compression test (UCT) are implemented to find out the compressive strength of samples. Blends selected from the dry side of the compaction curve own larger compressive strength than the others, which means an appropriate amount of water is enough for blends mixing. The results of laboratory tests are utilized to establish neural network models to predict engineering properties such as, compressive strength, optimum moisture content, and maximum dry density of treated marine clay. The correlation between the additive contents as well as curing days and compressive strength has been approved according to the test results. The properties such as maximum dry density, water contents, and additive contents are used as inputs to predict the compressive strengths. The predicted consequences of neural networks are well fitted with laboratory test results.

Keywords: Compaction test · Neural network · Compressive strength · Marine clay

1 Introduction

Marine clay is a typical soil that can be found in coastal areas around the world. It is generally considered as problematic soil since the low strength of marine clay, makes it unsuitable for any engineering activity in its natural state, Geotechnical engineers are constantly striving to improve properties such as soil behavior, consolidation properties, and strength to meet major development needs and to meet specific requirements.

B. Fatahi et al. (Eds.): GeoChina 2021, SUCI, pp. 1–14, 2021.
https://doi.org/10.1007/978-3-030-79854-3_1

Prefabricated vertical drains (PVDs) with preload are used to improve the geotechnical properties of soft marine clay which was used to support the subgrade of a highway at Mumbai, India (Sarika et al. 2015). The feasibility of treated marine clay slurries that possess adequate strength for heavy burden activities used as valid landfill material was studied (Wu et al. 2015). Conventional methods such as vacuum preloading are limited in the bearing capacity available. Chemical physics methods are effective when the initial water content is at a high level. It has been reported that using geotextile bags under simulated conditions can significantly improve the consolidation rate. The project team of the Hong Kong Housing Authority developed a green treated method to handle marine clay for In-situ backfilling. The successful cement stabilization process is accomplished with the optimal mix ratio determined. A complete mix of 80% of wet marine clay and 5% premixed Portland cement added with 15% In-situ granular material maximizes engineering properties, cost-benefit and environmental worth of the mixed material (HKHA 2016).

A series of laboratory tests have been conducted to verify the feasibility of pure crushed glass used for reclamation and earthworks (Sunny et al. 2014). Using green waste, such as glass cullet, is suggested for the application of bedding and backfilling for the trench because it can be used locally and also greener and more environmentally friendly (CWC 1993).

The stabilization of sensitive clay with quick lime was studied in the laboratory by (Locat et al. 1990). They found that sensitive clay treated with quick lime achieved significant levels of strength with sufficient curing time. A study on the marine clay stabilized by the lime-cement combination was conducted by (Venkateswarlu et al. 2014). The effectiveness of inorganic by-products to stabilize the clay has been reported, the compressive strength of clays stabilized with by-product binders generally exceeds the strength in reference tests with cement (Heikki 2000). All the above results from various pieces of literature can prove that the stabilization of clay with certain recycled materials is of practical application.

The neural network, namely feedforward multilayer perceptron, is one of the most significant types of supervised neural networks (Singh et al. 2001). The conception of the perceptron is historical and means the function executed by the nodes. Feedforward signifies that there are certain inputs and outputs, and the data flows in one direction. A feedforward neural network is shown in Fig. 1.

Five different artificial neural networks were established to predict compaction parameters using 126 results from the standard compaction test (Günaydın 2009). Soil fraction, specific gravity, soil type, liquid limit, and plastic limit are described as inputs to map onto outputs, which are optimum moisture content and maximum dry density. Neural network has the advantages over traditional digital computing techniques and symbolic processing, since they can adapt to subtle changes in the nature of problems and tolerate errors of input data. The neural networks can also be transferred easily between computing systems, the whole benefits are compatible with civil engineering applications.

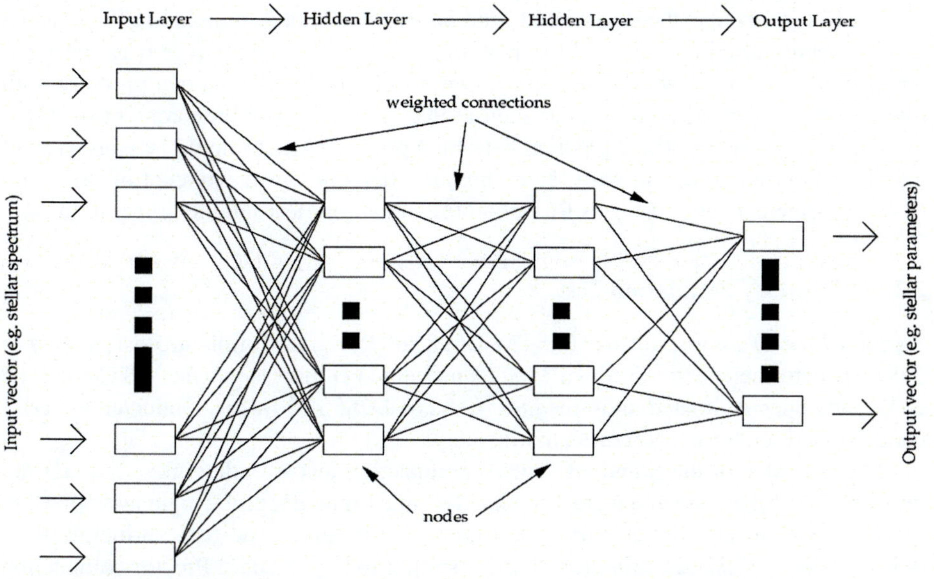

Fig. 1. Feedforward artificial neural network architecture (Singh et al. 2001)

2 Materials and Laboratory Test

2.1 Preparation of Testing Materials

In this study, the natural moisture content of Macau marine clay varies from 50% to 60%. To prepare the test samples with remolded water content, the following procedure was carried out:

The Large pieces of marine clay are removed from the bucket and placed on a metal tray. The marine clay together with metal tray were placed in an oven set at a condition of 60 °C for 24 h. Then a hammer or a grinder is used to crush the large pieces of clay into powders, and sieved by NO. 10 (2 mm) sieve. The soil less than 2 mm will be sealed and stored for later use. The moisture content of the oven-dry soil is around 3%.

Since the natural moisture content of glass cullet is almost zero, no drying is required. The original glass cullet is also sieved by NO. 10 (2 mm) sieve, the powders less than 2 mm will be stored for later use. The Macau marine clay mixes with glass cullet with different ratios will be the material for sample preparation.

Besides, cement and lime are both chosen as additives to treat marine clay. The cement used in the laboratory is Portland cement bought from the cement factory of Macau. The main chemical components of Portland cement are calcium oxide (CaO), silica (SiO_2), ferric oxide (Fe_2O_3), and aluminum oxide (Al_2O_3). The main minerals of Portland cement are tricalcium silicate (C_3S), dicalcium silicate (C_2S), tricalcium aluminate (C_3A), and tetra calcium iron aluminate (C_4AF). The cement is preserved in plastic bags and kept in the dark and dry condition. Adding different percent of cement to the clay soil is regarded as the blends for laboratory test sample preparation.

The lime utilized in the laboratory is purchased from Zhuhai, mainland China, which is an air-hardening inorganic gelling material with calcium oxide (CaO) as the main component. Lime is made of limestone, dolomite, chalk, shells, and other products with high calcium carbonate content, calcined at 900–1100 °C. The lime content (CaO) is over 98% in the light of the ingredient list of the product, which should be conserved in a dark and dry environment with sealed bottles. Blending diverse percent of lime with Macau marine clay are materials for upcoming laboratory test sample preparation.

2.2 Laboratory Evaluation Test

Standard Proctor compaction test (ASTM-D698 2007), Harvard miniature test (Scavuzzo 1984), penetrometer test (Cordova 1957), unconfined compression test (ASTM-D2166 2013), and unconsolidated undrained test (ASTM-D2850 2007) are conducted to verify the enhanced properties of treated marine clay.

The objective of the Standard Proctor compaction test is to determine the relation between dry unit weight and water content. The maximum dry unit weight and optimum moisture content are determined by the compaction curve. The basic principles and objectives of the Harvard miniature test are similar to the Standard Proctor compaction test. However, less material is required for the conduct of the Harvard miniature test. The size of samples yielded by the miniature test is more appropriate for the unconsolidated-undrained test, which is aimed at measuring the shear strength of prepared materials.

Geotechnical engineers use pocket penetrometers to acquire the shear strength and related unconfined compressive strengths quickly and approximately. The test samples are obtained from the Standard Proctor compaction test, the estimated unconfined compressive strengths of samples are directly given when the 6.4 mm dia. penetration plunger was used. To acquire the unconfined shear strength of compacted soil samples, the unconfined compression test is adopted. The unconfined shear strength of the soil approximates its undrained shear strength.

3 Established Artificial Neural Network Embedded in MATLAB

3.1 Model for OMC and Maximum Dry Unit Weight

87 compaction test results have been utilized to build up the neural network model. In the study, the whole data are collected from laboratory test and previous literature. Liquid limit (LL), plastic limit (PL), cement content (CC), lime content (LC), and compaction energy (CE) are parameters of inputs in the neural network model, the optimum moisture content (OMC) and maximum dry unit weight (MDUW) of Macau marine clay are parameters of outputs. The data range of OMC and MDUW is shown in Table 1. The basic architecture of the back-propagation feedforward neural network is displayed in Fig. 2.

Table 1. Range of inputs and outputs values for OMC and MDUW modeling

Inputs	Max	Min
Liquid limit (%)	91	25
Plastic limit (%)	38.1	14
Cement content (%)	16	0
Lime content (%)	12	0
Compaction energy (kJ/m^3)	2700	558.31
Outputs	Max	Min
Optimum moisture content (%)	36	6.5
Maximum dry unit weight (kN/m^3)	22.148	12.5

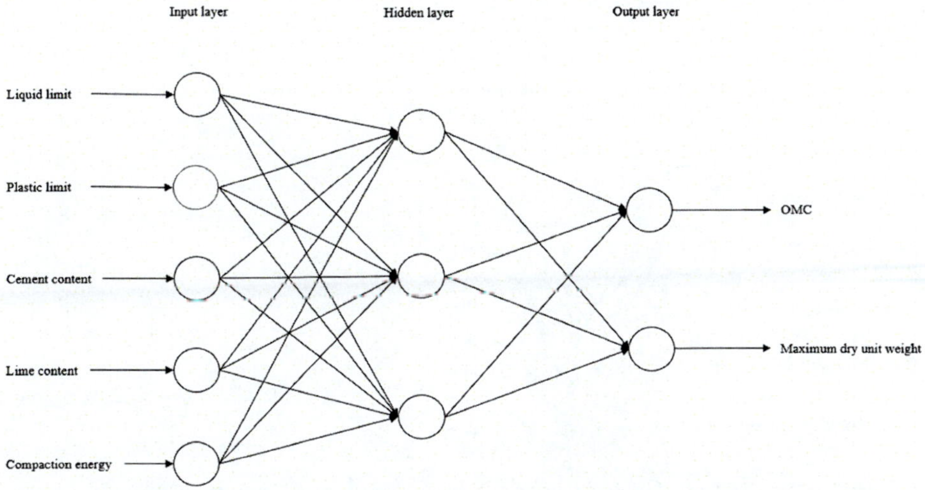

Fig. 2. The architecture of the model for estimating OMC and MDUW

3.2 Model for Compressive Strength

265 test results of UC test and UU test have been utilized to build up the neural network model. Liquidity index (LI), cement content (CC), lime content (LC), dry unit weight (DWU), optimum moisture content (OMC), curing days (CD), confining pressure (CP), and compaction energy (CE) are parameters of inputs in the neural network model. The unconfined compressive strength is regarded as the output. Table 2 shows the data range of compressive strength. The architecture of the model for estimating compressive strength is shown in Fig. 3.

Table 2. Range of inputs and outputs values for strength modeling

Inputs	Max	Min
Liquidity index (%)	0.4	−1.44
Cement content (%)	16	0
Lime content (%)	12	0
Dry unit weight (kN/m^3)	18.73	11.98
Optimum moisture content (%)	35	12.75
Curing day	90	0
Confining pressure (kpa)	404	0
Compaction energy (kJ/m^3)	2700	558.31
Outputs	Max	Min
Compressive strength	5610	31.3

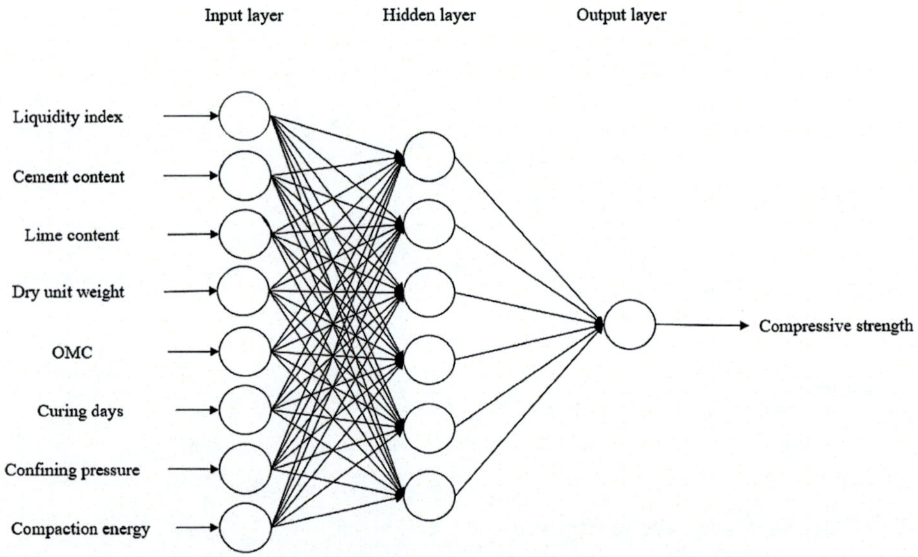

Fig. 3. The architecture of the model for compressive strength

4 Results and Discussion

4.1 Laboratory Test Results of Treated Marine Clay Blends

Figure 4, 5, and 6 display the compaction properties of treated marine clay with different additives in different ratios.

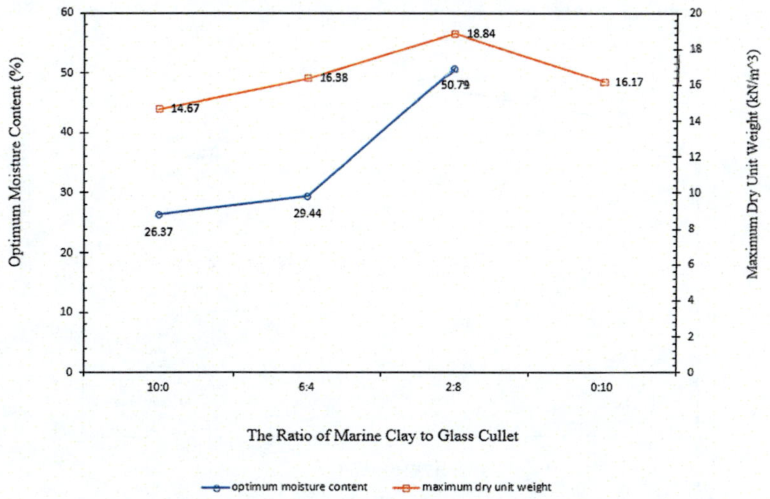

Fig. 4. Compaction properties of glass-treated clay in different ratio

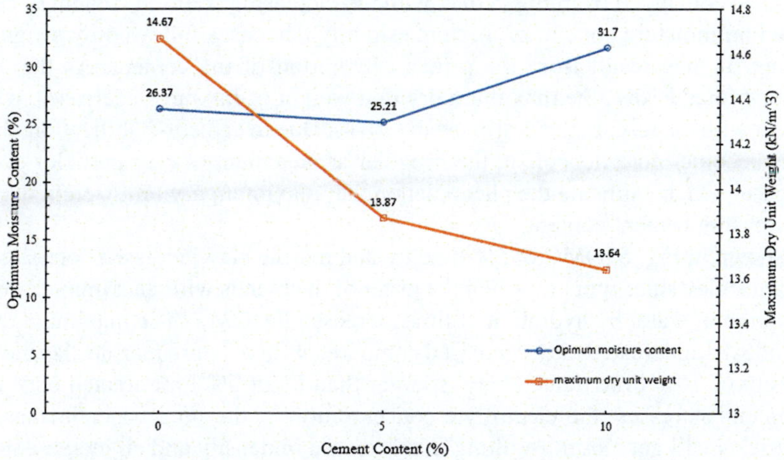

Fig. 5. Compaction properties of cement-treated clay in different ratio

From Fig. 4, with the increasing addition content of glass cullet, the overall tendency of the optimum moisture content of mixtures is rising, which causes that much more water is needed for the lubrication of glass cullet-marine clay blends. The maximum dry unit weight is also increasing until the blends have no content of marine clay, it is the reason that glass cullet has a greater density than pure marine clay.

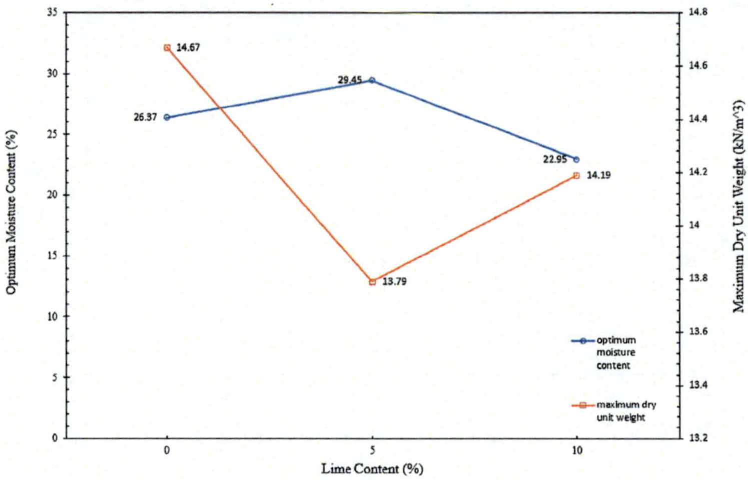

Fig. 6. Compaction properties of lime-treated clay in different ratio

It can be concluded from Fig. 5 that, with the increasing addition content of cement, the optimum moisture content of mixtures slightly decreases followed by a significant increasing. It may result from the primary hydration of the cement. As the cement content is 5% and 10%, the maximum dry unit weight of mixtures decreases, which is consistent with the research of Karim et al. (2014). Due to cement-stabilization, the clay particles become more flocculent, the irregular arrangement of clay particles occupies more space, which explicate the phenomenon that maximum dry unit weight decreases with increasing cement content.

As shown in Fig. 6, OMC of 5% lime treated marine clay increased compared with pure marine clay since quicklime and the water of the blends will react immediately, the consumption of water by hydration of lime increased the OMC. The maximum dry unit weight of 5% lime treated clay decreased compared with the pure marine clay, however, the density of 10% lime treated clay is larger than its of 5% lime treated clay. As the lime content increases, the electrolyte concentration of the pore water increases, the clay particles will get closer, resulting in a decrease in density and an increase in water content. With the further increase of lime, the attraction is offset by the hydration and penetration forces, resulting in the overall force changing from attraction to repulsion (Monowar Hussain 2015). In different soils, a certain lime content occurs at the inflection point. As for the study, 10% lime has crossed the inflection point and dry density has begun to increase.

Figure 7 shows the variation of compressive strength of 10% cement-treated marine clay blends tested by the UU test with curing days. Figure 8 displays the compressive strength versus curing days of 10% lime-treated marine clay mixtures.

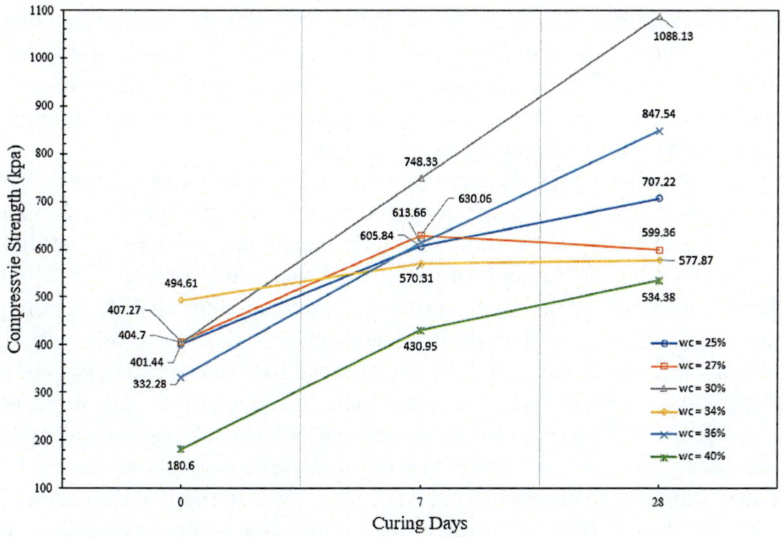

Fig. 7. UU strength versus curing days of 10% cement mixtures

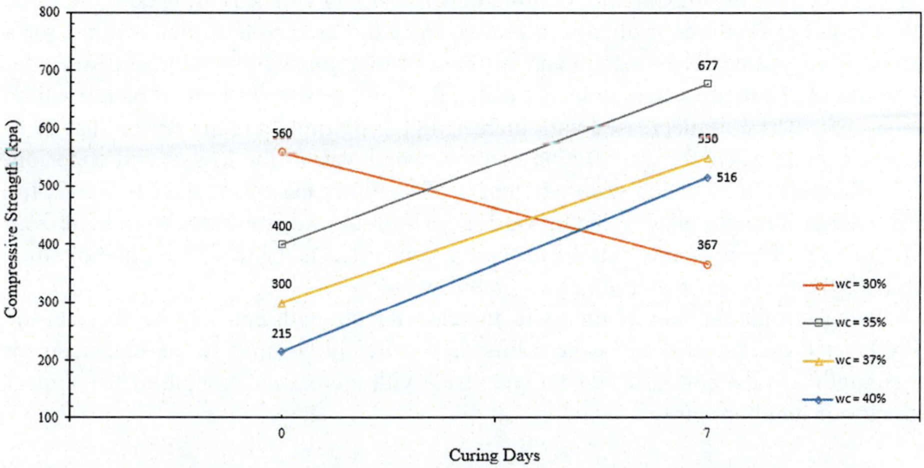

Fig. 8. UU strength versus curing days of 10% lime mixtures

The specimen is placed in the sealing bag to minimize the effect of water loss and promote cement hydration process. And then the sample will be preserved in water curing bath for 0,7 and 28 curing days to realize the chemical interaction between soil and cement, as the cement can attain its 90% strength on 28 days. From Fig. 7, it can be concluded that the increasing curing days can significantly improve the UU strength of mixtures. Since the hydration of cement and consequent pozzolanic reaction have relatively slow reaction rates, the formation of cementing products is slow either. It takes a long time for cement-treated clay to reach its maximum strength. The bond between the particles has been strengthened by cementing substance formed over time, which is the reason that the compressive strength of mixtures can rise. It is noticed that the compressive strength of 7 curing days is larger than its of 28 curing days for which water content is 27%, it may be caused by the inappropriate curing condition controlling, the drying ambient leads to the dropping of strength.

Apart from the curing days, the water content of the specimen does have an influence on the UU strength. Under the same curing days, the strength first increases and then decreases with increasing water content, which means that there is an optimal water content for improving the strength of blends. As the water content is 30% which is approaching to the optimum moisture content in compaction test of 10% cement treated marine clay, the compressive strength of blends are greater than the left with different water content on 7 curing days and 28 curing days. Too little water is not enough for the consumption of cement hydration, otherwise, too much water also has a negative effect on the bond strength provided by cementing substance (Ampera and Aydogmust 2005). Laboratory studies have shown that the cementing degradation caused by mean effective stress and deviatoric strain affects the peak shear strength of the cement-treated clay, resulting in the significant removal of the cementation bond, thus reaching the peak strength (Nguyen et al. 2017).

According to Fig. 8, the specimen with 10% lime content, 35% water content, and 7 curing days have the maximum UU strength which is 677 kPa. As curing days increase, the strength of mixtures gradually increases, since the hydration of lime can last for a relatively long time. The strength and stiffness of mixtures will be improved since the clay flocculated by lime over time. Specially, the compressive strength of blends which water content is 30% decreased with increasing curing time, it is the reason that there is no adequate water for the reaction of lime, which causes the loss of soil cohesion. Except the specimen which water content is 30%, either the curing days is 0 or 7, the compressive strength displays an upward trend with the water content decreased from 40% to 35%. The reason is that the excessive water affects the utilization rate of silica which is the main raw material in lime stabilization.

Increasing the amount of lime will increase the strength until all the silica in the clay is used up, the effect of excess lime is practically negative to the enhancement of strength. On the contrary, cement stabilized with increasing strength of the cement dosage constantly improved.

4.2 Prediction Results of Model for Estimating Compressive Strength

A certain quantities values of liquidity index, cement content, lime content, dry unit weight of mixtures, optimum moisture content, curing days, confining pressure, and compaction energy are used to predict the compressive strength of mixtures.

The cement content and lime content are changeable inputs in the application of the model, the distribution of the input of cement content is shown in Fig. 9 and the distribution of the input of lime content is shown in Fig. 10.

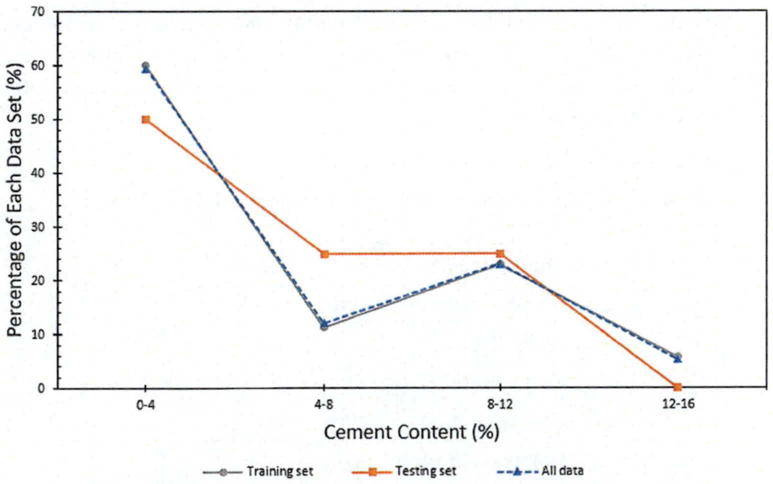

Fig. 9. Distribution of the input of cement content in the database

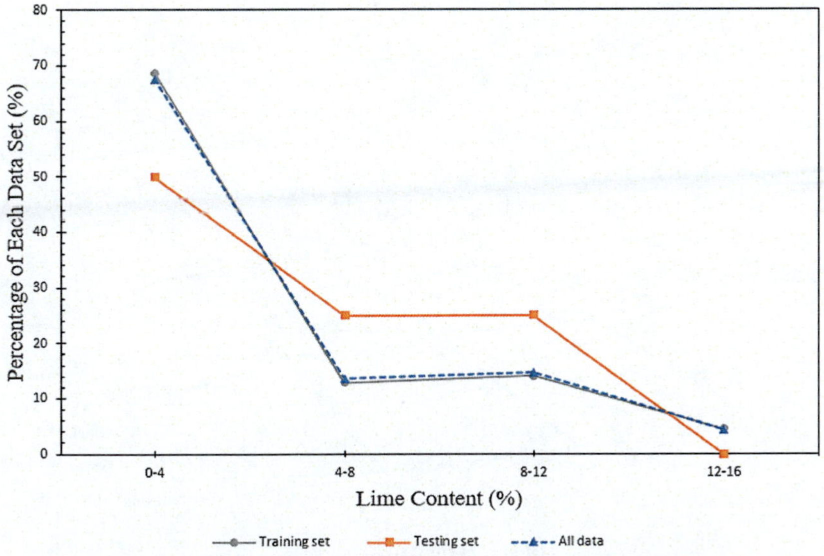

Fig. 10. Distribution of the input of lime content in the database

From Fig. 9, it can be concluded that the distribution of cement content in the training set and testing set are both approximately uniform. The distribution lines of the training set and testing set are highly consistent as shown in Fig. 10. The biases of the database in the prediction process are within a controllable range.

The values of inputs except for cement content and lime content are controlled variables during the modeling process, the specific values of these inputs are listed in Table 3. The predicted compressive strengths of blends that are mixed with both cement and lime which are displayed in Fig. 11.

Table 3. Values of inputs except cement content and lime content

Liquidity index	0.1667
DUW of mixture (kN/m^3)	13.5
OMC (%)	35
Curing duration (days)	7
Confining pressure (kPa)	0
Compaction energy (kJ/m^3)	558.31

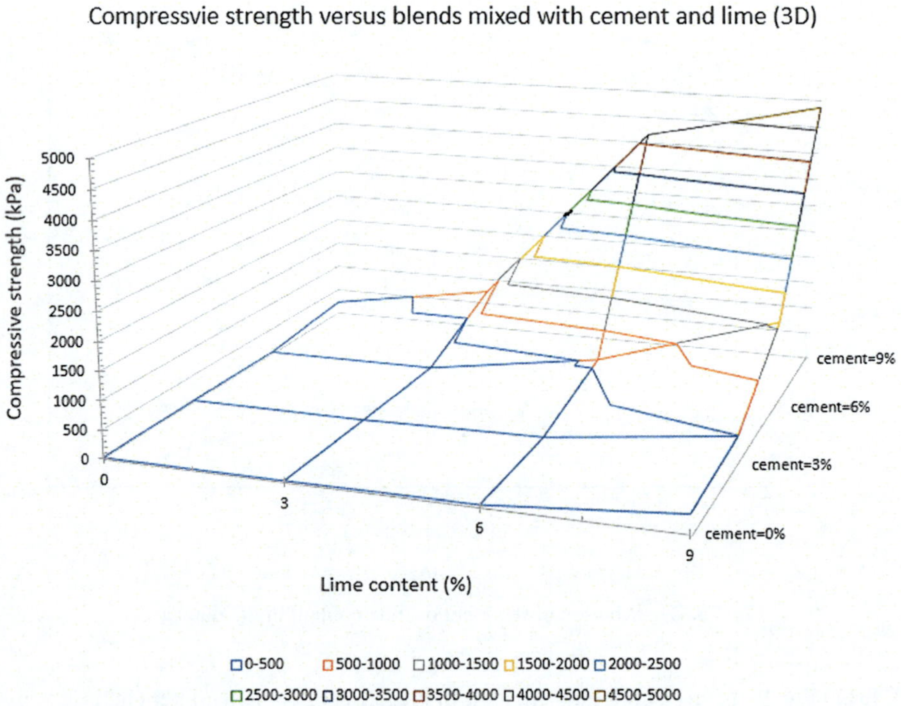

Fig. 11. Compressive strength versus blends mixed with cement and lime (3D)

According to the figure, with the additional content of cement and lime, the compressive strength of blends keeps increasing instantly. When the lime content of blends is at a high level, namely 9% lime content, the compressive strength of blends can be significantly improved with the increasing content of the cement.

5 Conclusion

The conclusion in the thesis is divided into two main parts: one is based on the results of a series of laboratory tests, which focuses on the compaction properties and compressive strength of treated Macau marine clay blends. The enhanced effect of various treatments such as mixing with glass cullet, cement, or lime will be summarized. Another part is about the establishing, determining, and evaluating of two artificial neural network models. Model I used several soil properties and test properties to predict the optimum moisture content and maximum dry unit weight of the blends. Model II is designed for predicting the enhanced compressive strength of treated Macau marine clay.

With the addition of glass cullet, the optimum moisture content and maximum dry unit weight of Macau marine clay increasing. The compressive strength of glass treated marine clay is larger than that of pure marine clay since the existence of adhesive strength between the surface of marine clay and glass cullet increased. With the increasing content of cement, the optimum moisture content of mixtures slightly decreases followed by a significant increasing. On the contrary, the maximum dry unit weight of mixtures is decreasing as the content of cement increases. The optimum moisture content of 5% lime treated marine clay increases compared with pure marine clay. The maximum dry unit weight of 5% lime treated marine clay decreased in comparison to pure marine clay since the density of lime-treated clays is less than pure marine clay.

Liquid limit, plastic limit, cement content, lime content, and compaction energy are effective parameters of inputs to forecast the optimum moisture content and maximum dry unit weight of Macau marine clay treated with cement or lime content. Liquidity index, cement content, lime content, dry unit weight of mixtures, optimum moisture content, curing days, confining pressure, and compaction energy are effective parameters of inputs to predict the compressive strength of Macau marine clay treated with cement or lime content.

References

Ampera, B., Aydogmust, T.: Recent experiences with cement and lime-stabilization of local typical poor cohesive soil. Geotechnik-kolloquium Freiberg (2005)

ASTM-D2166: Standard Test Method for Unconfined Compressive Strength of Cohesive Soil. ASTM International, West Conshohocken (2013)

ASTM-D2850: Standard Test Method for Unconsolidated-Undrained Triaxial Compression Test on Cohesive Soils. ASTM International, West Conshohocken (2007)

ASTM-D698: Standard Test Methods for Laboratory Compaction. ASTM International, West Conshohocken (2007)

Cordova, F.J.: Field application of the pocket penetrometer. Am. Soc. Testing Mater. Spec. Tech. Publ. **232**, 372–375 (1957)

CWC: Glass Feedstock Evaluation Project, Testing Program Design. Clean Washington Center, Dames and Moore Inc., West Conshohocken (1993)

Günaydın, O.: Estimation of soil compaction parameters by using statistical analyses and artificial neural networks. Environ. Geol. **57**, 203–215 (2009)

Heikki, K.: Stabilization of clay with inorganic by-products. J. Mater. Civ. Eng. **12**(4), 307–309 (2000)

Jacques, L., Marc-Andre, B., Marc, C.: Laboratory investigation on the lime stabilization of sensitive clays shear strength development. Can. Geotech. J. **27**, 294–304 (1990)

Karim, N.A.A., Rashid, A.S.A., Noor, N.M., Yaacob, H.: Effect of cement stabilized kaolin subgrade on strength properties. J. Appl. Sci. **14**(8), 842–845 (2014)

Monowar Hussain, S.K.: The influence of lime on the compaction behaviour of soils. J. Environ. Geotech. (2015)

Nguyen, L., Fatahi, B., Khabbaz, H.: Development of a constitutive model to predict the behavior of cement-treated clay during cementation degradation: C3 model. Int. J. Geomech. **17**(7) (2017)

Sarika, S.D., Deepankar, C., Salunkhe, J.S.: Ground improvement of marine clay for highway construction. In: IFCEE 2015 and ASCE 2015, Mumbai, India (2015)

Scavuzzo, R.: Use of the Harvard miniature apparatus for obtaining moisture unit weight relationships of soils (1984)

Singh, H.P., Bailer-Jones, C.A.L., Gupta, R.: An introduction to artificial neural networks. Autom. Data Anal. Astron. **I**, 51–68 (2001)

So, T.C., Lee, M.K., Hui, T., Shiu, Y.K.: Study of using recycled glass cullet as an engineering fill in reclamation and earthworks in Hong Kong. The Government of the Hong Kong Special Administrative Region (2014)

Sunny, S., Vogel, N., Howell, C., Vu, T.L., Aizenberg, J.: Lubricant-infused nanoparticulate coatings assembled by layer-by layer deposition. Adv. Funct. Mater. **24**, 6658–6667 (2014)

Venkateswarlu, D., Anjan-kumar, M., Prasada-raju, G.V.R.: A study on the lime – cement stabilized marine clay. J. Microbiol. Biotech. Environ Sci. **16**(2), 439–444 (2014)

Wu, D.Q., Xu, W.Y., Tjuar, R.: Improvements of marine clay slurries using chemical-physical combined method (CPCM). J. Rock Mech. Geotech. Eng. **7**, 220–225 (2015)

Prediction of Seismic Site Coefficients for Northeast Arkansas (NEA)

Md Rafiue Islam[1] and Zahid Hossain[2](✉)

[1] Arkansas State University, PO Box 1740, State University, Jonesboro, AR 72467, USA
mdrafiue.islam@smail.astate.edu
[2] Civil Engineering, Arkansas State University, PO Box 1740, State University, Jonesboro, AR 72467, USA
mhossain@astate.edu

Abstract. About 25% of bridges in Craighead County, a part of Northeast Arkansas (NEA) in the United States of America, are in a poor condition. The NEA region is situated within a major seismic zine called the New Madrid Seismic Zone (NMSZ). In recent years, several new bridges have been proposed for new highways or improvement of existing highways within this zone. With the existing vulnerability due to the presence of the nearby active fault and higher seismic design force requirements, future investment in the repairment of bridges and new construction are critical issues. The main objectives of this study are to estimate seismic site coefficients for this zone by ground response analysis and compare them with coefficients recommended by researchers and agencies such as the American Association of State Highway Official (AASHTO). The shear wave velocity profiles (SWVP) of this zone have been collected from historical construction projects in the NEA and Mississippi Embayment (ME) areas. The seismic hazard analysis (SHA) has been carried out for different locations within this zone. Using a unified 1-D (one-dimension) equivalent linear (EQL) and nonlinear (NL) site response analyses platform called DEEPSOIL (Version 7.0), a series of ground response analyses has been carried out to evaluate possible variabilities of SWVPs, dynamic soil properties, and SHA results. Finally, seismic acceleration site coefficients have been estimated for selected locations within this zone. The findings of this study are expected to help the city and state highway design engineers to calculate design forces for existing and new bridges that will lead to the avoidance of expensive field investigations.

Keywords: Earthquake · Site coefficient · Seismic hazard · DEEPSOIL · Shear wave velocity · Ground response analysis

1 Introduction

According to the United States Census Bureau, the population of the City of Jonesboro, a part of North-east Arkansas (NEA) was 75,866 in 2017, which was rising at a substantial rate. The Arkansas Office of Emergency Services along with several research studies has predicted several seismic-hazard (earthquake) scenarios, which are alarming for the

B. Fatahi et al. (Eds.): GeoChina 2021, SUCI, pp. 15–36, 2021.
https://doi.org/10.1007/978-3-030-79854-3_2

people of this city and surrounding areas. The Northeast Arkansas (NEA) region is very vulnerable in the event of a major earthquake due to its proximity to the New Madrid seismic zone (NMSZ). Within the eastern part of the United States, the NMSZ is considered the most seismically active zone. It has experienced some of the major earthquakes (e.g., Mw > 7.0 in the winter of 1811–1812) in the United States and has the potential of producing high magnitudes of earthquakes in the future. Additionally, the NEA lies within the upper Mississippi Embayment (ME), which has deep and unconsolidated sedimentary deposits of approximately 1 km. The ME consists of sediments of clay, silt, sand, and gravel; it has lignite ranging in age from Cretaceous to the recent Holocene (Hashash and Park 2004). This deep deposition imposes challenges in the estimation of local site effects during the propagation of ground motion. An idealized cross-section of the ME is given in Fig. 1.

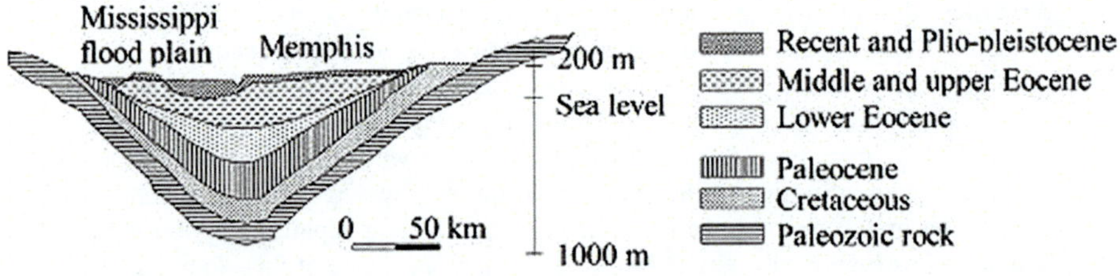

Fig. 1. Idealized cross-section of the Mississippi Embayment (Hashash and Park 2004).

Currently, the Arkansas Department of Transportation (ARDOT) has a plan to construct several highway bridges within NEA, and the ARDOT engineers usually follow the general guidelines recommended by the American Association of State Highway and Transportation Officials (AASHTO) on bridge design. Generally, the code-based design procedures such as the AASHTO guidelines do not account for the impact of deep deposited soft soil down to bedrock. During the propagation of earthquake ground motions in this identical geologic formation, short-period spectral acceleration may be attenuated and long-period spectral acceleration may be amplificated while performing the site-specific ground motion response analysis (SSGMRA). These scenarios cannot be captured by following code-based design procedures. With this existing vulnerability due to the presence of the nearby active fault and deep deposition of soft soil overlying the bedrock, the future investment in the construction or repairment of bridges has become a critical issue. Considering the amplification or deamplification imposed by deep deposited soil down to bedrock, the AASHTO suggests that a short period (01 to 0.5 s) structure may be over-designed with significant cost, and a long-period structure may be under-designed with significant risks. Hence, the AASHTO recommends performing the SSGMRA for the areas having deep deposited subsoil to ensure a safe and cost-effective bridge design by considering the better estimation of ground motion (AASHTO 2011).

Typically, a bridge with a pier height of more than 115 ft (35 m) is considered a tall-pier bridge. The slenderness ratio of a tall-pier bridge is significantly higher than that of a medium or short pier bridge, thus the former may experience more instability and displacement problems at the top of the piers. The seismic behavior of tall-pier bridges is different from that of short-pier bridges because of the increased effects of higher modes and axial load due to seismic events. Based on the type of materials (pre-stressed concrete, cast-in-situ concrete, composite structures/concrete slab on steel beams, etc.) used in bridge constructions, the span length of short-period structures usually varies from 40–120 ft (12–36 m). Short-period bridges are generally constructed in the NEA region, which is mostly rural.

To comprehend the effect of deep deposited soil down to bedrock during the event of an earthquake, a study was conducted at three Missouri bridge sites by performing the SSGMRA (Rogers et al. 2007). It predicted surface site response amplification from 6 to 9 times higher than that of bedrock. It also indicated a serious failure of foundation in the event of a large earthquake over Mw 6.5 to 6.6 (Rogers et al. 2007). According to a few other researchers (Romero and Rix 2001; Hashash et al. 2019), there is a large impact on the transfer of bedrock motions to surface ground motions during a large earthquake while considering the deep deposited subsoil. According to Cox et al. (2012), the deep deposited soil down to bedrock within the NEA is expected to damp out during an event of a large earthquake. Therefore, it is very essential to perform the SSGMRA considering the deep deposited soil in the NEA zone.

To perform the SSGMRA, in this study, Equivalent Linear (EQL) and Nonlinear (NL) methods have been followed using a one-dimensional (1-D) site response analysis platform called DEEPSOIL 7.0. Finally, a delineated design response spectrum (DDRS) has been developed for each of the two selected sites (Bay and Marked Tree). The DDRS will assist ARDOT engineers to calculate the amplification factor or spectral acceleration for a short period as well as for a long period without any field investigation or other time-consuming methods.

2 Objectives

The main objectives of this study are:

- Collect Shear Wave Velocity (Vs) Profile (SWVP) data from the previously completed projects within the NEA and the ME areas.
- Gather routine geotechnical properties (unit weight, overconsolidation ratio, angle of friction, plasticity index, etc.) of subsoils.
- Collect and adjust appropriate input earthquake acceleration time histories based upon the target spectrum.
- Simulate the propagation of input ground motions from bedrock to the ground surface using a site response analysis platform called DEEPSOIL Version 7.0
- Estimate seismic site factors by developing the DDRS.
- Compare the estimated site coefficients with previously recommended ones.

3 Methodology

According to Matasovic and Hashash (2012), three general approaches are available for seismic site response evaluation of soil. They are: (1) the attenuation relationship approach, (2) the code-factor approach, and (3) the site response analysis approach. In this study, the site response analysis approach was followed. This same approach was followed in determining the seismic site coefficient for Monnet, Arkansas (Wood et al. 2019), and I-40 Replacement Bridge over the White River (Cox 2014). The basic inputs of this analysis are (1) various properties of soil i.e. unit weight, shear wave velocity profile, angle of friction, etc. (2) input ground motion time histories, and (3) simulation of wave propagation using various site response analysis platforms such as DEEPSOIL 7.0, SHAKE2000/D-MOD2000, etc. Then develop a delineated design spectrum based upon the site response analysis results. A flow diagram of the research methodology of the current study is shown in Fig. 2.

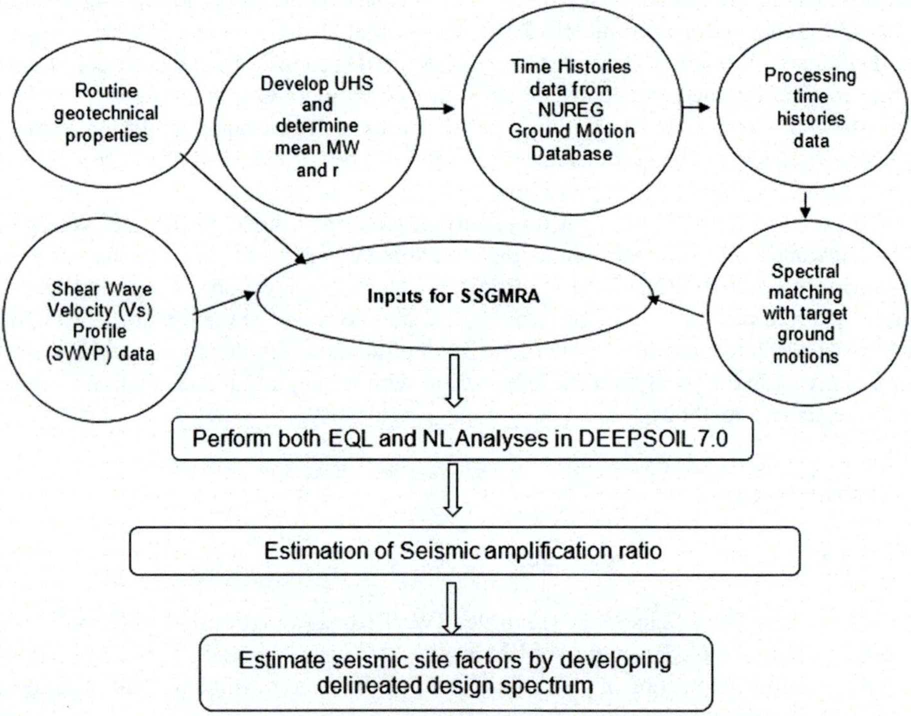

Fig. 2. Flow chart of research methodology

3.1 Collection of Various Properties of Soil

The EQL and NL methods were followed to develop mean ground surface spectral acceleration (SA) for Bay and Marked Tree. They are two of the ARDOT bridge sites in NEA. The locations of these ARDOT bridge sites are shown in Fig. 3. The median shear wave velocity (Vs) profiles of these sites were collected from a recently completed ARDOT project (Wood et al. 2019). The shear wave velocity profiles of these two ARDOT Bridge sites have been given in Table 1 (Wood et al. 2019). The upper 30 m median shear wave velocity V_{S30} values for these sites have been given in Table 2

(Wood et al. 2019). According to Table C3.10.3.1-1 (AASHTO 2017), these sites can be considered as Site Class D since the shear wave velocity ranging from 600 ft/s–1200 ft/s. Guidelines provided in Section 3.4.3.2 of the AASHTO Guide Specifications for LRFD Seismic Bridge Design (2nd Edition) (AASHTO 2011) were followed to perform the SSGMRA. In this study, medium sand was considered along the entire soil column based upon shear wave velocity profile and geologic deposition of these project locations (Wood et al. 2019). The friction angle was considered 30° and there was no cohesion of soil. The plasticity index (PI) was assumed as zero and the OCR was assumed as one. Based upon the geologic deposition, the unit weight of soil for these sites was ranging from 1700 kg/m^3–2100 kg/m^3.

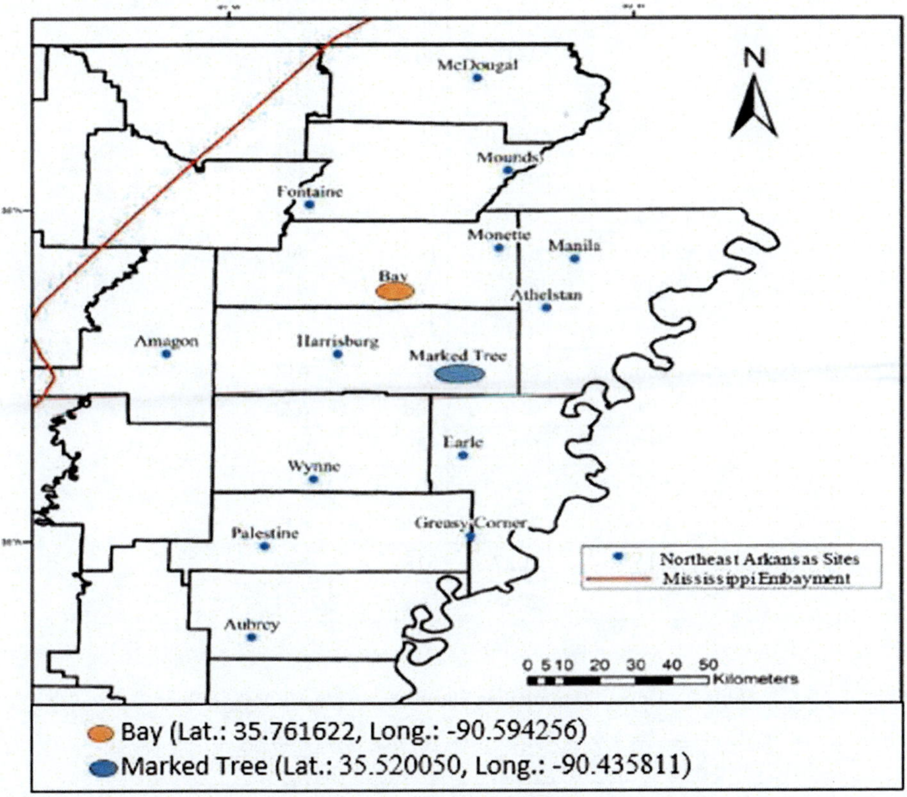

Fig. 3. Dynamic site characterization testing locations in Northeast Arkansas (Wood et al. 2019).

Since these soil parameters are not available to the team, they have been assumed based on information available in the literature. These assumptions were also made by other researchers on multiple projects. For instance, Cox (2014) used these assumptions to estimate seismic site factors for I-40 Replacement Bridge over the White River. Besides, Wood et al. (2019) used these assumptions to estimate seismic site factors for Monnet, Arkansas.

Table 1. Shear wave velocity profiles (SWVP) for Marked Tree and Bay.

Marked Tree		Bay	
Depth (m)	Vs (m/s)	Depth (m)	Vs (m/s)
0	120	0	131
1.6	136	1.7	175
4.5	166	3.8	209
7.5	217	8	222
12	252	19	232
21	287	31	295
33	354	38	454
47	441	49	527
61	508	65	556
49	561	78	573
104	590	96	584
421	608	117	593
193	614	148	599
241	626	173	605
295	637	212	611
343	643	255	614
401	656	300	620
455	676	349	625
541	704	403	631
641	769	451	643
715	1037	518	669
853	2068	593	2284

Table 2. Upper 30 m median shear wave velocity V_{S30} values.

Site name	Latitude	Longitude	VS30 (ft/s)
Bay	35.761622	−90.594256	692.08
Marked Tree	35.520050	−90.435811	698.64

3.2 Input Ground Motion Time Histories

3.2.1 Uniform Hazard Spectrum (UHS) and De-aggregation

The USGS Unified Hazard tool was used to perform de-aggregation analysis and to develop a uniform hazard spectrum (UHS). The selected edition of this hazard tool was conterminous U.S. 2014 (update) (v.4.2.0). From the de-aggregation analysis, the mean magnitude of an earthquake and mean source (earthquake) to site distance for these sites were determined. The summary of de-aggregation has been given in Table 3. The uniform hazard spectrum (UHS) was chosen for Site Class A (2000 m/s). The return period was 1000 years, or the probability of exceedance was about 5% in 50 years. This UHS was chosen as the design target spectrum, and it provided the ground motion (g) for the spectral period ranging from 0 to 2 s. A linear interpolation technique was followed in developing the UHS. The UHS for each of the two sites is shown in Fig. 4. The spectral period was plotted in log scale and the value ranged from 0.01 to 2.0 s.

Table 3. Summary of deaggregation using USGS unified hazard tool.

Site name	Latitude	Longitude	Mean magnitude of earthquake, M_W	Mean source (earthquake) to site distance, r (km)
Bay	35.761622	−90.594256	7.49	47.91
Marked Tree	35.520050	−90.435811	7.50	16.79

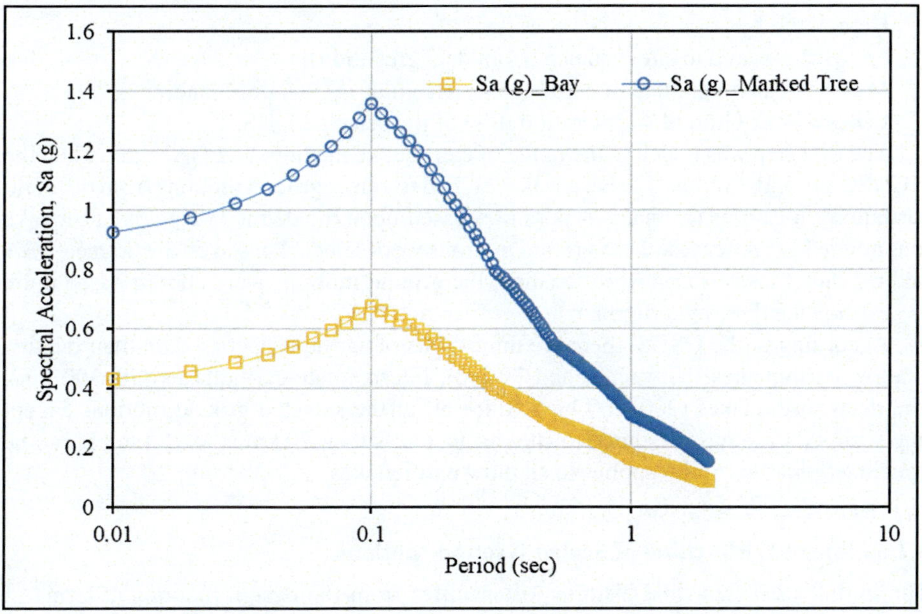

Fig. 4. Uniform hazard spectrum (UHS) as target spectrum.

Each value of the UHS curve originates from different hazard curves that have the same design hazard level (i.e., a return period of 1000 years or the probability of exceedance is 5% in 50 years). From Fig. 4, it is seen that Marked Tree has higher target ground motion (i.e., higher spectral coordinates) than that of Bay for the given probability of exceedance. This implies that the seismic design force for designing structures bridges will be higher for Marked Tree compared to that of Bay. Besides, for performing seismic hazard analysis, higher ground motions (i.e. higher PGA, S_S, S_1) are required to be chosen for Marked Tree. Thus, the development of the UHS as target ground motion has a great impact on the estimation of overall seismic site coefficients.

3.2.2 Selection of Ground Motions

Twelve ground motions were selected from the US Nuclear Regulatory Commission Regulation (NUREG) database (McGuire et al. 2001) for the central and eastern United States. The summary of the selected ground motions is provided in Table 4. For selecting the ground motions, quantitative thumbs rules, defined in Cox et al. (2012) were followed. These thumbs rules are:

- Earthquake magnitude varies from (M_W – 1) to (M_W + 1).
- Range of source to site distance from 0.5 * r to 2.0 * r.
- The range of selecting PGA was from 0.5 * PGA to 2 * PGA (0.33 * PGA to 3 * PGA is acceptable if necessary).
- Select as many different earthquake locations as possible.
- Duration of earthquake > 10 s.

Here,
r = Mean source to site distance (from deaggregation),
M_W = Mean magnitude of earthquakes ((from deaggregation), and.
PGA = Peak Ground Acceleration (0 s) of the selected UHS.

The first step was to define the range of earthquake magnitude, PGA, and r. Then the NUREG ground motions database was searched to select ground motions from different earthquake locations (as many as possible) based upon the defined range of earthquake magnitude M_w. After that these ground motions were selected based on the defined PGA and r. When these criteria were satisfied, the ground motions were chosen based upon the earthquake duration criteria.

According to the USGS, there are three types of earthquakes based on their depths: shallow earthquakes (between 0 and 70 km), intermediate earthquakes (70–300 km), and deep earthquakes (300–700 km). Since all of the selected ground motions except one with 94.4 km have a source to site distance less than 70 km (r < 70 km), it can be concluded that this study applies to shallow earthquakes.

3.2.3 Spectral Matching of Scaled Ground Motions

The initially selected time histories often differ from the design motions in terms of shaking peak amplitude and response spectral ordinates. Due to these reasons, modification of time histories is required to be used in the analysis. Two modification methods

Table 4. Summary of the selected ground motions

Site name	Filename	Earthquake	PGA	Magnitude	Source to Site Distance, r	Duration
			(g)	Mw	(km)	(s)
Bay	SHL000	Cape Mendocino	0.648	7.1	33.8	14.4
	MDR000	Kocaeli, Turkey	0.201	7.4	34.6	15.5
	SOD285	San Fernando	0.228	6.6	58.1	9.1
	RAN000	Northridge	0.240	6.7	55.2	11.9
	SOR-UP	Northridge	0.263	6.7	54.1	11.5
	29P000	Landers	0.207	7.3	42.2	20.4
	IZN180	Kocaeli, Turkey	0.239	7.4	29.7	14.5
	HWA056-N	Chi-Chi, Taiwan	0.203	7.6	48.7	12.3
	TCU015-N	Chi-Chi, Taiwan	0.218	7.6	47.3	22.4
	ABY090	Landers	0.434	7.3	69.2	19.6
	FER-T1	Tabas, Iran	0.358	7.4	94.4	23.2
	TCU047-W	Chi-Chi, Taiwan	0.700	7.6	33.0	12.9
Marked Tree	BRN090	Loma Prieta	1.236	6.9	10.3	8.6
	TCU128-N	Chi-Chi, Taiwan	0.305	7.6	9.7	29.9
	IZT180	Kocaeli, Turkey	0.401	7.4	7.7	10.1
	DAY-TR	Tabas, Iran	0.947	7.4	17.0	9.7
	SHL000	Cape Mendocino	0.648	7.1	33.8	14.4
	ARC000	Kocaeli, Turkey	0.298	7.4	17.0	10.4
	TCU046-W	Chi-Chi, Taiwan	0.336	7.6	14.3	18.8
	TCU047-N	Chi-Chi, Taiwan	1.168	7.6	33.0	10.8
	TCU095-W	Chi-Chi, Taiwan	0.782	7.6	43.4	14.0
	TCU136-N	Chi-Chi, Taiwan	0.302	7.3	8.9	22.9
	FSD172	San Fernando	0.344	6.6	27.5	7.3
	BOL090	Duzce, Turkey	1.765	7.1	16.0	6.4

are commonly used in practice: (1) simple scaling approach, and (2) spectrum matching approach. In this study, the spectrum matching approach has been followed. Besides, the initially selected time histories were processed (if applicable) using the SeismoSignal 2020 tool, which is an earthquake software for signal processing of time histories.

The spectrum matching approach has been used. SeismoMatch 2020 was used to match the scaled ground motions in terms of response spectra. It is an earthquake software for adjusting the earthquake accelerograms to match a specific target response spectrum. Ground motions were matched with the target spectrum (UHS) at least from 0.5Tn–2.0Tn, as mentioned in Section 3.10.2 of the AASHTO Guide Specifications for LRFD

Seismic Bridge Design (AASHTO 2017). These matched ground motions have been used as input ground motion in the site response analysis using DEEPSOIL 7.0. A summary of the matched ground motions with the target spectrum has been given in Fig. 5.

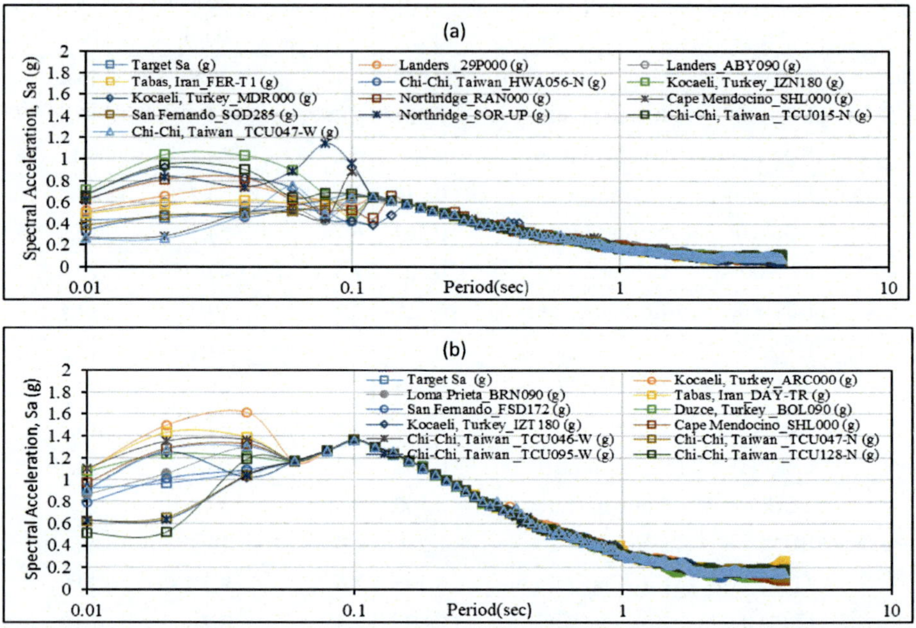

Fig. 5. Summary of matched ground motions with target spectrum for: (a) Bay and (b) Marked Tree.

Spectrally matched twelve ground motions with the target spectrum have been shown both for both Bay (Fig. 5a) and Marked Tree (Fig. 5b). These ground motions were chosen as input ground motions for performing the SSGMRA using DEEPSOIL 7.0. Initially selected twelve ground motion's spectral shape did not match with a target spectrum. There were variations in terms of (PGA, 0.2-s spectral acceleration, and 1.0_s spectral acceleration. After the spectral matching performed in SeismoMatch 2020, these twelve ground motions were selected because their spectral shape did match with the target spectrum. Matching of ground motions with the target motion was performed in such a way that the spectral shape of matched ground motion did not get much disturbed compared to its original ground motion as well as matched the shape of the target spectrum.

3.3 Simulation of Wave Propagation Using Site Response Analysis Platform

DEEPSOIL 7.0, a one-dimensional site response analysis platform capable of performing both EQL and NL analyses, was used to perform the seismic response analysis (Hashash et al. 2019). For performing both EQL and NL analyses, the General Quadratic/Hyperbolic Model (GQ/H) soil model was used along with Non-Mashing re/unloading formulation.

Darendeli (2001) proposed reference curves for both modulus reduction and damping to develop soil's dynamic properties such as [G^*/G_{max}] and material damping [D]. In this study, dynamic properties of soil were developed for each soil layer using Darendeli (2001) empirical relationships in DEEPSOIL 7.0. These reference curves were fitted using GQ/H (Groholski et al. 2016) soil model for each soil layer along with the MRDF-UIUC reduction factor fitting procedure (Philips and Hashash 2009; Hashash et al. 2016).

The dynamic property of soil is generally influenced by mean effective confining pressure, soil plasticity index (PI), and over consolidation ratio (OCR). In this study, medium sand was considered along the soil column based upon shear wave velocity and geologic deposition of this project location (Wood et al. 2019). Based upon the geologic deposition, the unit weight of soil for these sites was ranging from 1700 kg/m^3–2100 kg/m^3. The target shear strength of the nonlinear shear modulus reduction (G/G_{max}) is calculated using the Mohr-Coulomb equation (Eq. 1).

$$\tau_{target,} = cVs + \sigma_v' * \tan(\phi) \tag{1}$$

Where,

$\sigma v'$ = effective stress at the mid-depth of the interested soil layer,

ϕ = is the friction angle of the soil layer, and

cVs = judgement-based shear strength developed at 0.1% shear strain for a linear elastic material with 80% of the maximum shear modulus derived from the shear wave velocity (Vs) of the soil layer under consideration, as shown in Eq. 2 (DEEPSOIL 2019).

$$cVs = \rho * Vs^2 * 0.8 * 0.1\% \tag{2}$$

The coefficient of earth pressure at rest (K_0) was calculated as shown in Eq. 3.

$$K_0 = [1 - \sin(\phi)] * OCR^{\sin(\phi)} \tag{3}$$

3.4 Development of Delineated Design Spectrum

The final site-response spectra for these sites have been developed based upon a weighted average, a 50%–50% average, of both EQL and NL analyses. At first, an average of the amplification ratios for both EQL and NL analyses from DEEPSOIL were obtained. These amplification ratios were obtained by dividing the surface response by input rock responses. Then these average amplification ratios were multiplied by the AASHTO Site Class A spectrum (Wood et al. 2019). This AASHTO Site Class A spectrum was developed as per the AASHTO LRFD Bridge Design Specifications (Eight Edition) Sections 3.10.3.2, and 3.1.4.

The delineated design spectrum for all periods was determined as follows:

- Determine the AASHTO design spectrum for Site Class D as per the AASHTO LRFD Bridge Design Specifications (Eight Edition) Sections 3.10.3.2 and 3.1.4.
- Multiply the AASHTO design spectrum for Site Class D by 2/3.
- Multiply amplification ratios (50%–50% weighted average of both EQL and NL analyses) by AASHTO Site Class A spectrum.

- Plot these three spectra in a single plot.
- Check and make sure the delineated design spectrum is between (1) AASHTO design spectrum for Site Class D and 2/3 of AASHTO design spectrum for Site Class D. This means for developing the delineated design spectrum, the upper limit is regular (1) AASHTO design spectrum for Site Class D and the lower limit is two-third (2/3) of the AASHTO design spectrum for Site Class D.

The DDRS for the sites is shown in Figs. 9 and 13.

4 Results

4.1 Bay

The maximum shear strain was found to be 0.165% from both EQL and NL analyses. A summary of maximum shear strain for EQL and NL analyses is shown in Fig. 6. From

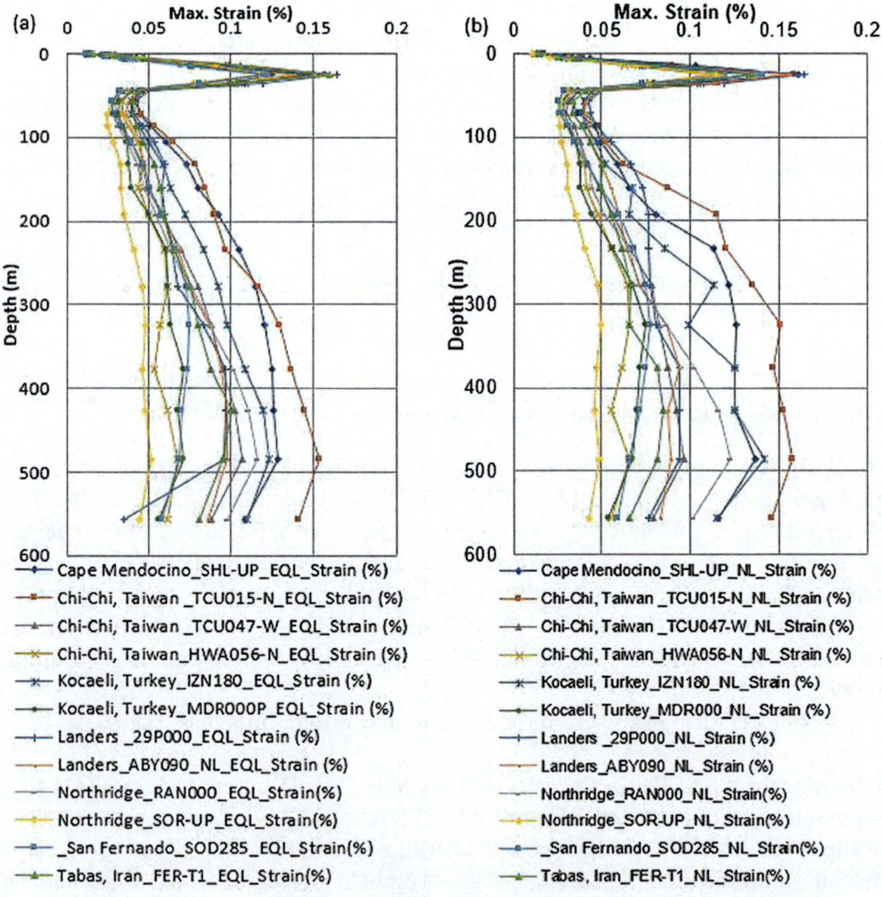

Fig. 6. Summary of maximum shear strain for Bay: (a) EQL analysis and (b) NL analysis.

this figure, it is found that maximum strain occurs at a depth of 484.5 m both for EQL and NL methods. In both types of analysis, the dominating ground motion was found to be the same (Chi-Chi, Taiwan). Sample comparisons between EQL and NL methods for each type of input rock motions are shown in Fig. 7.

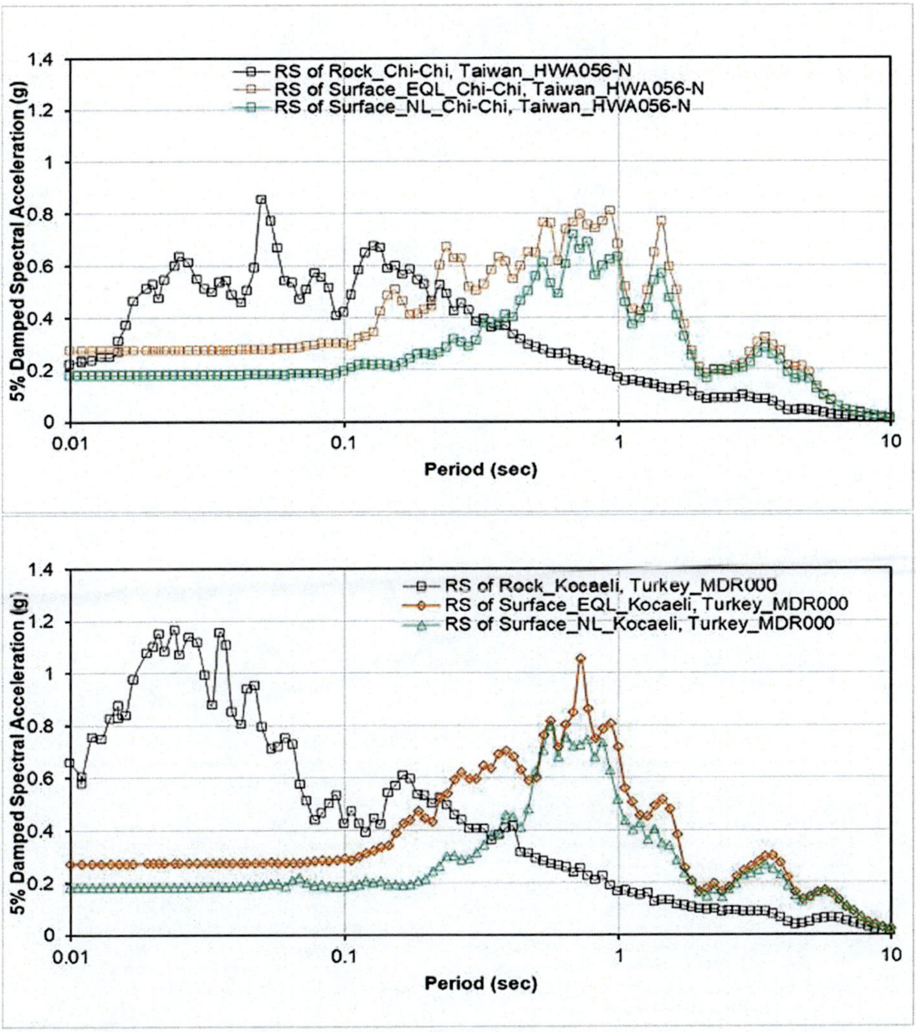

Fig. 7. Samples comparisons between EQL and NL analyses results for each type of input rock motion for Bay.

A summary of seismic site response spectra for both EQL and NL methods is shown in Fig. 8. In Figs. 9(a) and 9(b), the mean spectral accelerations (SA) both at surface and rock are shown for both EQL and NL analysis methods. Finally, the DDRS has been shown in Fig. 9(c).

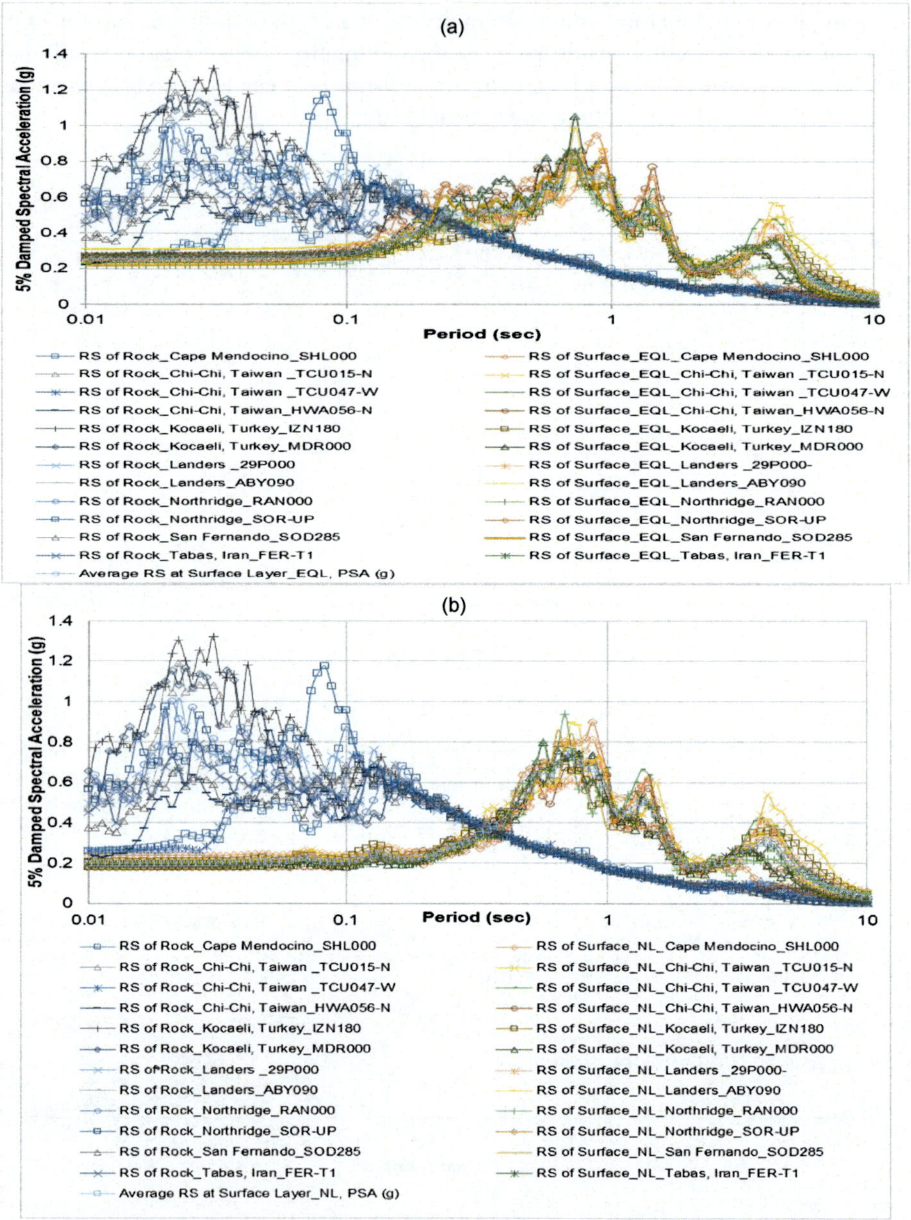

Fig. 8. Summary of response spectrum for Bay: (a) EQL analysis and (b) NL analysis.

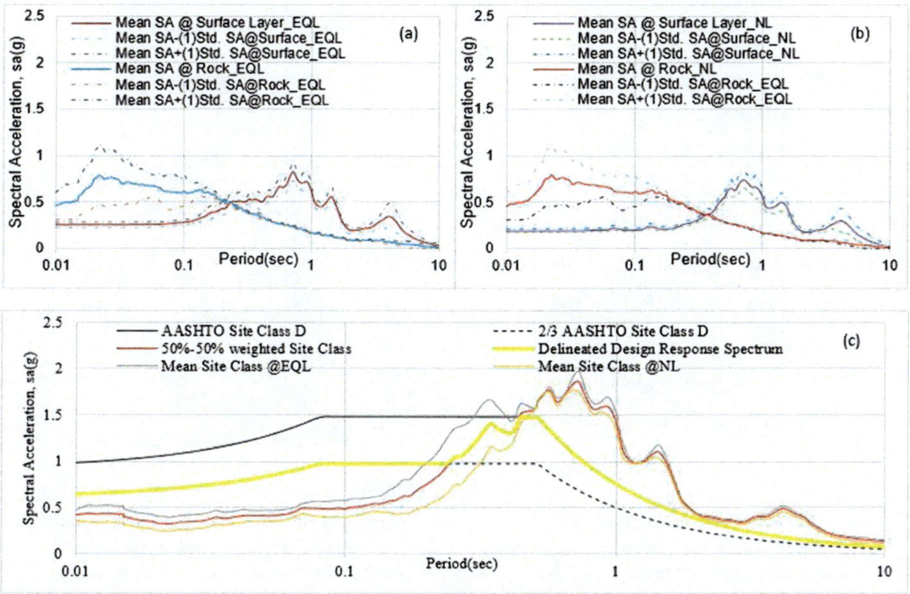

Fig. 9. Comparisons of SA at both surface and rock for Bay: (a) EQL analysis, (b) NL analysis, and (c) DDRS.

In Figs. 9(a) and 9(b), the SA along with standard deviation both at surface level and rock level have been shown for both EQL and NL analyses methods respectively. From these figures, it is seen that the mean SA at surface level exceeds the rock one at 0.253 s for the EQL analysis method whereas, for the NL analysis method, the mean SA at surface level exceeds the rock one at 0.345 s. Besides, in Fig. 9(c), the Mean Site Classes for both EQL and NL analysis methods have been shown along with their 50%–50% weighted Site Class. To represent the impact of both EQL and NL analyses on the estimation of seismic site coefficient, a 50%–50% weighted factor was considered during the development of the DDRS. From this Fig. 9(c), it is seen that the DDRS shows a reduction of 33% on PGA (0.01 s) compared to that of AASHTO. This same reduction continues up to approximately 0.238 s. From 0.238 s to approximately 0.443 s, the DDSR remains between the 2/3 and the regular (1) AASHTO general procedure response spectrum. From 0.443 s to the rest of the periods, the spectrum is greater than that of the AASHTO.

4.2 Marked Tree

The maximum shear strain from the EQL analysis was found to be 0.330% and 0.342% from NL analyses. A summary of maximum shear strain for EQL (a) and NL (b) has been shown in Fig. 10. From this figure, it is found that maximum strain occurs at a depth of 428 m both for EQL and NL analyses. In both analyses, the dominating ground motion was found to be the same as Chi-Chi, Taiwan. Sample comparisons between EQL and NL methods for each type of input rock motion have been shown in Fig. 11.

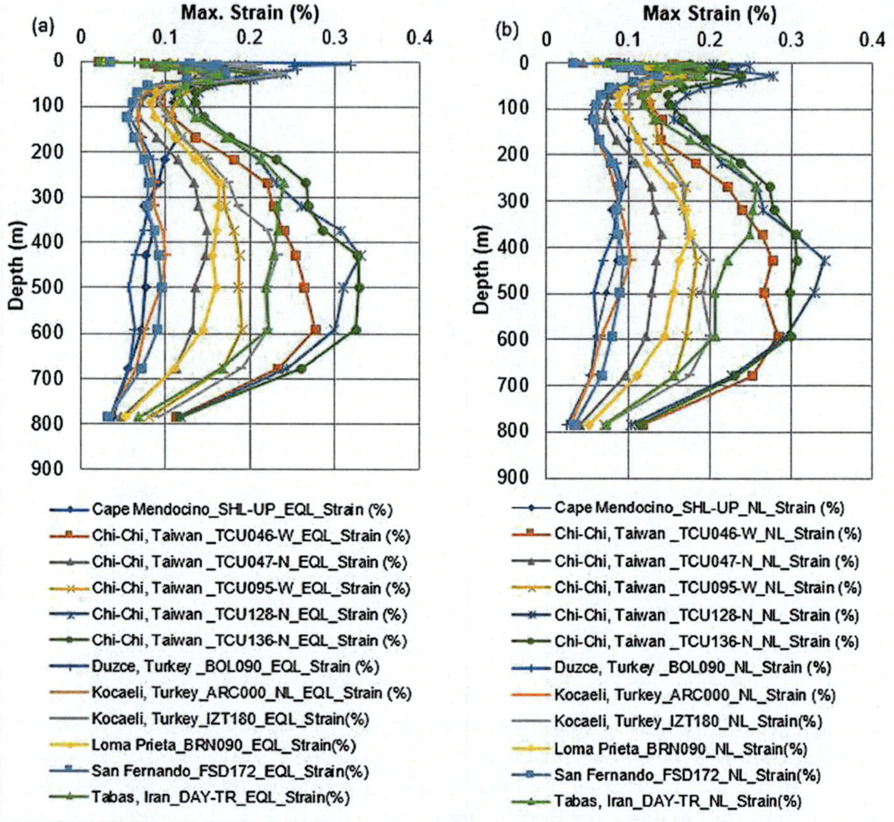

Fig. 10. Summary of maximum shear strain for Marked Tree: (a) EQL analysis and (b) NL analysis.

A summary of seismic site response spectra for both EQL and NL methods is given in Fig. 12. In Figs. 13(a) and 13(b), the SA at both surface and rock have been shown for both EQL and NL analysis methods. Finally, the DDRS has been shown in Fig. 13(c).

In Figs. 13(a) and 13(b), the SA along with standard deviation both at surface level and rock level have been shown for both EQL and NL analyses methods respectively. From these figures it is seen that mean SA at surface level exceeds the rock one at approximately 0.305 s for the EQL analysis method whereas, for the NL analysis method, mean SA at surface level exceeds the rock one at approximately 0.416 s. Besides, in Fig. 13(c), Mean Site Classes for both EQL and NL analyses have been shown along with 50%–50% weighted Site Class. From this Fig. 13(c), it is seen that Mean Site Class @ EQL shows earlier amplification and higher SA than that of NL. It is also seen that DDRS shows a reduction of 33% on PGA (0.01 s) compared to that of AASHTO. This same

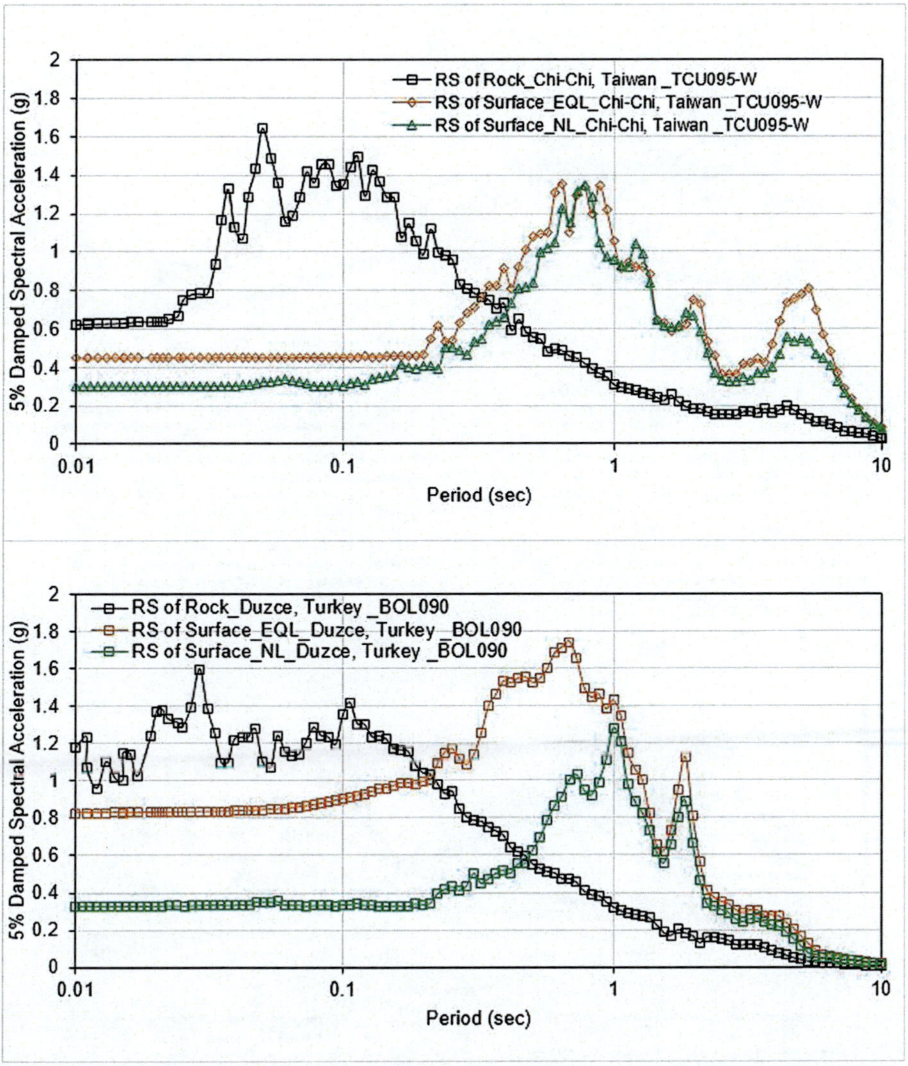

Fig. 11. Samples comparisons between EQL and NL analyses results for each type of input rock motion for Marked Tree.

reduction continues up to approximately 0.287 s. From 0.287 s to approximately 0.471 s, the DDSR remains between the 2/3 and the regular (1) AASHTO general procedure response spectrum. From 0.443 to the rest of the periods, the spectrum is greater than that of the AASHTO.

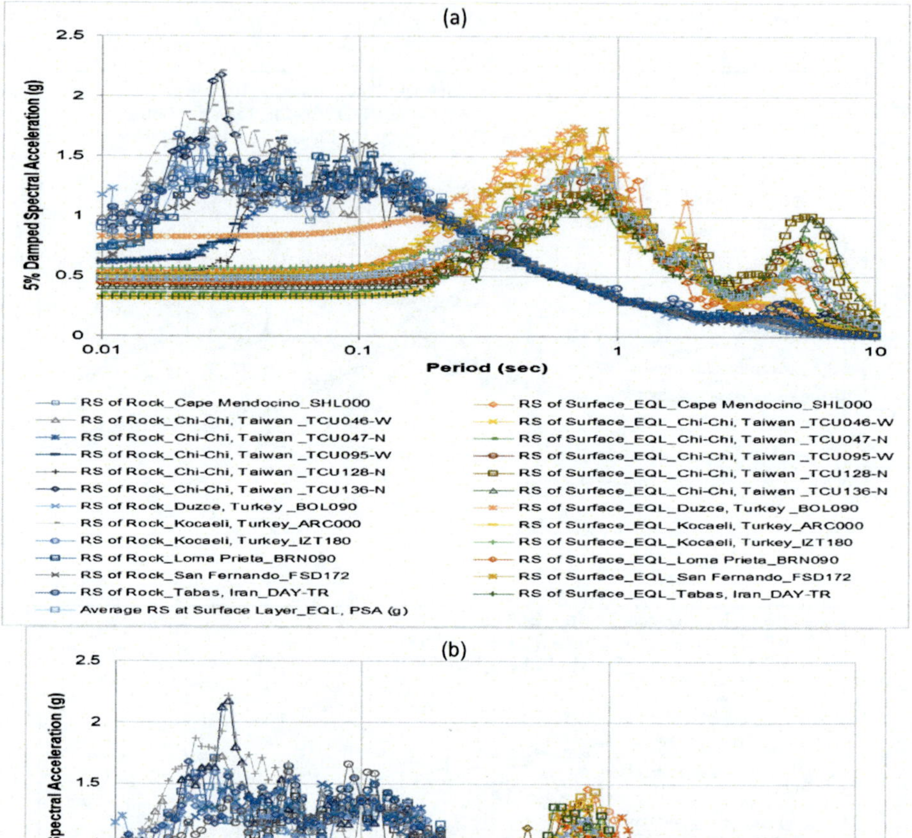

Fig. 12. Summary of response spectrum for Marked Tree: (a) EQL analysis and (b) NL analysis.

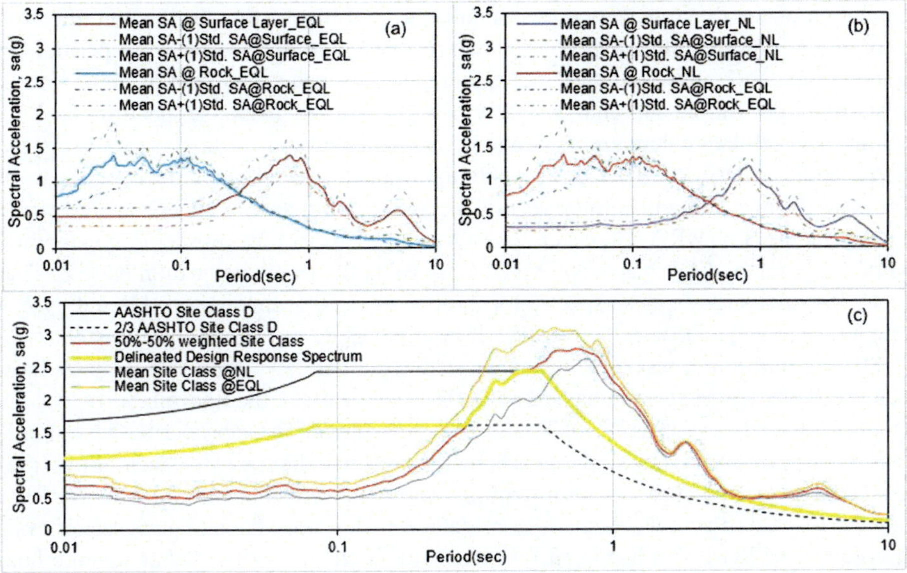

Fig. 13. Comparisons of SA at surface and rock for Marked Tree: (a) EQL analysis, (b) NL analysis, and (c) DDRS.

5 Discussion

In this study, seismic hazard analysis has been carried out for two ARDOT bridge locations (Bay and Marked Tree) using both EQL and NL methods. From the seismic hazard analysis on Bay, it has been found that there exist differences between EQL and NL analysis results. The EQL analysis shows higher responses in most of the cases at periods less than 1 s as well as at periods higher than 1 s than that of NL (Figs. 7 and 8). This pattern is also observed on the seismic hazard analysis for Marked Tree (Figs. 11 and 12). The EQL also shows an earlier amplification than that of the NL analysis for Bay, as seen in Figs. 9(a) and 9(b). Similar observations have also been made for Marked Tree, as seen from Figs. 13(a) and 13(b). The difference between EQL and NL analyses is well pronounced during the propagation of strong ground motions. In this study, it is seen that Marked Tree has the target motion compared to that of Bay. From the analysis results, it is seen that the gap between SA values of the EQL and NL analyses is greater for Marked Tree (Fig. 11) compared to that of Bay (Fig. 7). This is because a very flat response at high frequencies can be produced in the EQL analysis due to high damping values at sites where high shear strains are expected (Griffiths et al. 2016a). Besides, using constant stiffness and damping throughout the analysis (Hashash and Park 2004), the EQL might show higher responses both at short (<1 s) and long (>1 s) periods. On the other side, the soil behavior under large strains from strong ground motions at soft soil sites can be predicted in a better way in the NL analysis, since it accounts for changes in soil properties at each time step (Kim et al. 2016).

From these analysis results, it can be concluded that the EQL overestimates the seismic site amplification both at short and long periods. Because of its simplicity in the calculation of the dynamic properties of soil, the EQL overestimates the amplification factor and might give higher responses than that of input motions for short period. On the other hand, NL shows a lower surface response spectrum than that of input motion for short period. But the NL analysis shows a higher surface response spectrum for long periods. The NL can calculate the dynamic properties of soil for large strain using the GQ/H model in DEEPSOIL 7.0. It counts the impact of the largely deposited subsoil of a zone. Hence, the entire thickness of the soil column should be considered while performing the SSGMRA. However, the EQL analysis is still the most common method in practice, and the EQL has been proved to be valuable for studies within NMSZ. That is why the final site-response spectra for these sites have been developed based upon a weighted average, a 50%–50% average, of both the EQL and NL analyses results.

Staying within the scope of this study, data of previously completed projects have been used to perform seismic hazard analysis. In the case of the unavailability of data, some soil properties such as PI, OCR, and ϕ were assumed. These properties have significant impacts on soil dynamic properties behavior. Since there exists a lack of observational data on seismic history for this region, these data were collected from the NUREG database instead of actual seismic history data. Besides, only the site response analysis platform DEEPSOIL 7.0 has been followed in this study. Various other seismic hazard analysis platforms such as D-MOD2000 (Non-Linear Analysis), and SHAKE2000 (Equivalent Linear Analysis) can be used to perform the seismic analysis. Taking all of these scenarios into account, there might be a significant impact on the seismic hazard analysis presented here. Hence, further study is required to make a more concrete statement on seismic site coefficients reported in this study.

6 Conclusions

The purpose of this study is to bring out the usefulness of performing site-specific ground motion response analyses (SSGMRA) over the code-based design spectrum for the seismic design of structures like bridges within Northeast Arkansas (NEA) as a means of reducing short-period design ground motions. In this study, SSGMRA has been performed in a total of two ARDOT bridge sites. The code-based design procedure does not count the impact of deep soft soil down to bedrock such as the geologic formation of NEA. As a result, the code-based design spectrum may lead to over-designing the short-period structures at significant cost and under-designing the long-period structures at significant risk. Most of the bridges of ARDOT have fundamental periods in the range of 0.1–0.5 s and this fundamental period usually falls into the "short period" range where SSGMRA can reduce the anticipated seismic design forces compared to that of the AASHTO. On the other hand, for long-period structures, the SSGMRA may predict amplification of anticipated seismic design forces compared to that of the AASHTO.

The estimated spectral acceleration for Bay was found to be 33% less than that of the AASHTO bridge design guideline for a period of 0.01 s to 0.238 s. For Marked Tree, it was 0.01 s to 0.287 s. Hence, it can be concluded that the seismic design force can be reduced by 33% for these ARDOT bridge sites for the above-mentioned period. This leads to reducing the cost of short period bridge structures. On the other hand, the estimated spectral acceleration for Bay was found to be greater than that of AASHTO after 0.443 s. And, for Marked Tree, it was found to be 0.502 s. Hence, it can be concluded that the SSGMRA predicted amplification of seismic design force compared to that of AASHO after the above-mentioned period. Therefore, a significant financial saving can be achieved while designing short period structures (i.e. short span bridges) for these ARDOT bridge sites for the above-mentioned short periods.

References

Wood., C. M., Baker, E., Deschenes, M., Himel A.K.: Deep Shear Wave Velocity Profiling in Northeastern Arkansas, TRC1603 (2019)

American Association of State Highway and Transportation Officials (AASHTO): Guide Specifications for LRFD Seismic Bridge Design, 1st edn., AASHTO, Washington, D.C. (2009)

American Association of State Highway and Transportation Officials (AASHTO): Guide Specifications for LRFD Seismic Bridge Design," 2nd ed. Washington, D.C. (2011)

American Association of State Highway and Transportation Officials (AASHTO): Guide Specifications for LRFD Seismic Bridge Design," 8th ed. Washington, D.C. (2017)

Cox, B.R.: Site-Specific Ground Motion Response Analysis Results for the I-40 Replacement Bridge over the White River (AHTD Job BB0610) (2014)

Guerreiro, P., Kontoe, S, Taborda, D.: Comparative Study of Stiffness Reduction and Damping Curves (2012)

Griffiths, S.C., Cox, B.R., Rathje, E.M.: Challenges associated with site response analyses for soft soils subjected to high-intensity input ground motions. Soil Dyn. Earthq. Eng. **85**, 1–10 (2016a)

Griffiths, S.C., Cox, B.R., Rathje, E.M., Teague, D.P.: Mapping dispersion misfit and uncertainty in vs profiles to variability in site response estimates. J. Geotech. Geoenviron. Eng. **142**(11), 1–12 (2016b). 04016062

Groholski, D.R., Hashash, Y.M.A, Kim, B., Musgrove, M., Harmon, J., Stewart, J.: Simplified model for small-strain nonlinearity and strength in 1D seismic site response analysis. J. Geotech. Geoenviron. Eng. **142**(9), 1–14 (2016). 04016042

Hashash, Y.M.A., Musgrove, M.I., Harmon, J.A., Groholski, D.R., Phillips, C.A., and Park, D.: DEEPSOIL 7.0, User Manual (2019)

Hashash, M.A., Park, D.: Evaluation of seismic site factors in the Mississippi Embayment. I. Estimation of dynamic properties. Soil Dyn. Earthq. Eng. **25**(2005), 133–144 (2004)

Phillips, C., Hashash, Y.M.A.: Damping formulation for nonlinear 1D site response analyses. Soil Dyn. Earthq. Eng. **29**(7), 1143–1158 (2009). https://doi.org/10.1016/j.soildyn.2009.01.004

NEHRP Consultants Joint Venture: Selecting and Scaling Earthquake Ground Motions for Performing Response-History Analyses, NIST GCR 11-917-15 (2011)

Kim, B., et al.: Relative differences between nonlinear and equivalent-linear 1-D site response analyses. Earthq. Spectra **32**(3), 1845–1865 (2016)

McGuire, R.K., Silva, W.J., Costantino, C.J.: "Technical Basis for Revision of Regulatory Guidance on Design Ground Motions: Hazard- and Risk- consistent Ground Motion Spectra Guidelines, NUREG/CR – 6728 (2001). https://www.nrc.gov/docs/ML0131/ML013100012.pdf

Rogers, D.J., Karadeniz, D., Kaibel, C.K.: Seismic site response modeling for three Missouri river highway bridges. J. Earthquake Eng. **11**, 400–424 (2007)

Rathje, E.M., Kottke, A.R., Trent, W.L.: Influence of input motion and site property variabilities on seismic site response analysis. J. Geotech. Geoenviron. Eng. **136**(4), 607–619 (2010)

Liu, W., Stephenson, W.: Seismic response of two specific sites in the new Madrid seismic zone. In: Geotechnical Engineering for Transportation Projects (GeoTrans). Los Angeles, California, pp. 1682–1690 (2004)

Kramer, S.L., Arduino, P., Sideras, S.S.: Earthquake Ground Motion Selection, Washington State Transportation Center (2012). https://www.wsdot.wa.gov/research/reports/fullreports/791.1.pdf

Cox, B.R., Ellis, T.B., Griffiths, S.C.: Site-specific seismic ground motion analyses for transportation infrastructure in the New Madrid seismic zone, Mack-Blackwell Rural Transportation Center (2012)

Darendeli, M.: Development of a new family of normalized modulus reduction and material damping curves. Ph.D. dissertation, Civil Engineering Dept., University of Texas, Austin, Texas (2001)

Hashash, M.A., Matasovic, N.: Practices and Procedures for Site-Specific Evaluations of Earthquake Ground Motions," National Cooperative Highway Research Program, NCHRP Synthesis 428 (2012)

Characterization and Computational Fluid Simulation of Conditioned Muck During EPB Shield Tunneling in Gravelly Sand Stratum

Jiazheng Zhong[1(✉)], Shuying Wang[1,2], Pengfei Liu[1], Haibo Wang[1], and Zhengri Liu[1]

[1] College of Civil Engineering, Central South University, Changsha 410075, China
jiazhengzhong@csu.edu.cn

[2] Key Laboratory of Engineering Structure of Heavy Haul Railway, Central South University, Ministry of Education, Changsha 410075, China

Abstract. Soil conditioning is an available auxiliary measure to improve the efficiency and safety of earth pressure balance (EPB) shield tunneling. In order to optimize soil conditioning strategy during tunneling in gravelly sand stratum, it is essential to understand rheological and compressive behavior of conditioned muck and its flow characteristics inside chamber. This paper establishes a constitutive model of compressible modified Bingham fluid for characterizing macroscopically rheological and compressive behavior of foam-conditioned gravelly sand and determines the model parameters by vane shear tests with variable shear rate and lateral confining compression tests. Based on the above theory and computational fluid dynamics (CFD), crucial spatial distribution characteristics of conditioned muck static pressure and flow speed inside chamber are captured by numerical simulations, which is verified the reliability in simulating muck movement during tunnelling. Overall, investigation of macro-hydrodynamics behavior of foam-conditioned gravelly sand and numerical simulation of muck movement in excavation chamber and screw conveyor will inspire soil conditioning assessment while EPB shield tunneling in gravelly sand stratum.

Keywords: Earth pressure balance shield · Soil conditioning · Compressible modified bingham fluid · Computational fluid dynamics · Distribution of static pressure and flow speed

1 Introduction

Earth Pressure Balance (EPB) shield method is widely used in urban subway tunnel construction for less disturbance to the ground. It is worth mentioning that the safety and high efficiency of EPB shield tunnelling depend on smoothness of muck inflow and outflow as well as stability of excavation face. In order to prevent above risks and ensure EPB shield tunnels safely and efficiently, muck is usually transformed into an ideal paste with a certain compressibility and a relatively low shear strength using conditioners (Wang et al. 2020).

B. Fatahi et al. (Eds.): GeoChina 2021, SUCI, pp. 37–48, 2021.
https://doi.org/10.1007/978-3-030-79854-3_3

In recent years, several researchers studied smoothness of muck inflow and outflow or stability of excavation face with numerical simulation of EPB system applied for overcoming limitations of in-site monitoring and laboratory scale model test. Firstly, analyse the mechanical behaviour of conditioned muck through laboratory soil test (Hu et al. 2020). Furthermore, put forward the theoretical model reflecting the complex mechanical characteristics of conditioned muck and calculate related parameters (Hu and Rostami 2020; Yang et al. 2020). Finally, establish numerical model to explore the rheology behaviour of conditioned muck, the interaction between shield components and muck as well as ground deformation induced by tunnelling (Meng 2012; Qu et al. 2019). In sum, this method economically reproduces EPB shield tunnelling process under different strata and accurately predicts muck mechanical behaviour or ground deformation.

For modelling EPB shield tunnelling, many researchers have been established CFD models of EPB shield to investigate the temporal and spatial distribution and interaction between muck flow speed and pressure. Lin (2013) treated muck as incompressible Hershel-Bulkley fluid and then established a CFD model of cutterhead and chamber in Harbin shield project. The influence of the cutterhead opening ratio and forms on muck pressure distribution in chamber has been discussed. However, it ignored the compressibility of muck and the interaction between the chamber and screw conveyor. Meng (2012) revealed pressure and flow speed distribution with simulating compressible non-Newtonian fluid like muck in the whole reduced scaled EPB shield, whose pressure or flow speed field is different from reality so that the corresponding simulation results can only be used for qualitative analysis. Dang and Meschke (2020) numerically investigated muck pressure spatial distribution in size-different chambers with various components, such as pressure imbalance between the left and right side and pressure fluctuations. Their model considered the compressibility of foam and pressure dependence of muck rheology, but not figured out the interaction between muck pressure and flow speed distribution.

In general, additional studies are underway to discuss a proper theoretical model for characterizing the complex mechanical behaviour of conditioned muck, analyse muck physical field characteristics in EPB shield using CFD. The authors tested the compressive and rheological behavior of foam-conditioned gravelly sand by pressurized vane shear apparatus, established compressive modified Bingham model to represent conditioned muck and then build a CFD model of muck movement inside EPB shield. The model was verified by comparing actual and simulated values of tunnelling parameters. Furthermore, the spatial and temporal distribution of muck pressure and flow speed in excavation face, chamber and screw conveyor and their interaction were analysed.

2 Rheological Test of Foam-Conditioned Gravelly Sand

2.1 Soil Sample and Apparatus

The soil sample was foam-conditioned gravelly sand obtained from tunnel project of Kunming Rail Transit. Figure 1 shows the grain size distributions of soil sample. Foam, generated by the dispersed foam agent and water were used to condition the excavated soil in front zone of cutterhead and inside chamber. Thus, the basic properties of foam and conditioned soil are shown in Table 1.

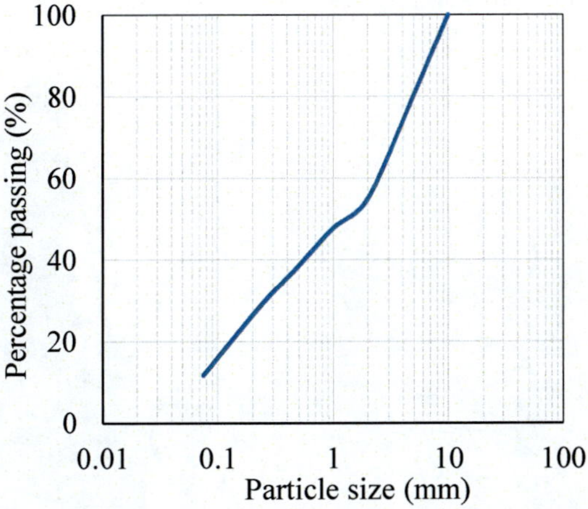

Fig. 1. Grain size distributions of soil sample

Table 1. Physical characteristics of soil sample

Foam	Foam injection ratio (FIR)		Foam expansion ratio (FER)	Half-life period
	19%		10	6 min
Foam-conditioned soil	Moisture content	Void ratio	Slump value	Expansion value
	22.5 %	1.15	7.1 cm	22.2 cm
	Internal friction angle	Cohesion	Density at atmospheric pressure	
	6°	2.47 kPa	2027.1 kg/m^3	
Slump of conditioned soil		 Oblique view		 Top view

Lateral compression test and vane shear test of foam-conditioned gravelly sand under undrained condition were carried out with the pressurized vane shear apparatus (see Fig. 2) proposed by Liu et al. (2020). The apparatus is composed of sample cavity, pressure device, shear device and sensors.

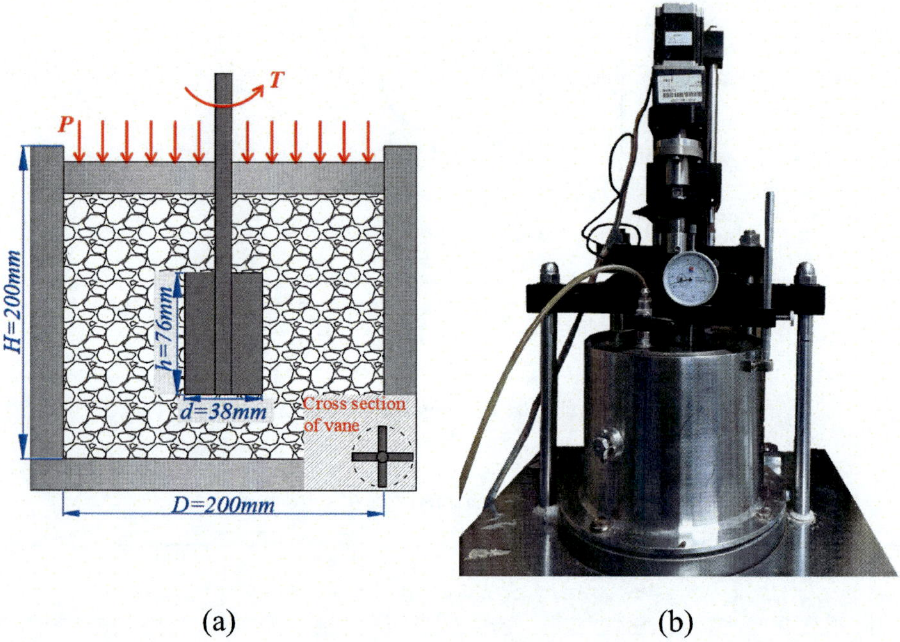

Fig. 2. Pressurized vane shear apparatus: (a) Schematic diagram; (b) Physical picture

The normal displacement of soil sample monitored by dial indicator represents the compression amount under the normal pressure in lateral compression test. Furthermore, vane torque and rotation speed obtained from vane shear tests can be transformed into shear stress and shear strain rate by Eq. (1) and (2), respectively (Meng and Li 2011).

$$\tau = \frac{6M}{\pi\left(d^3 + 3d^2h\right)} \tag{1}$$

$$\dot{\gamma} = \frac{\pi n}{5400} \times \frac{D^2}{D^2 - d^2} \tag{2}$$

where τ is shear stress, M is vane torque, D is diameter of cavity, d is diameter of vane, H is height of cavity, h is height of vane.

2.2 Case Setting

In order to obtain the compression and shear strength characteristics of conditioned muck under similar shield working environment, the normal pressure applied on the soil sample is set as 0, 50, 100, 150, 200, 250, 300 kPa while the vane rotation speed is set as 1/24, 1/12, 1/8, 1/6, 1/4, 1/3 rpm. Table 2 is the conversion table of vane speed and shear strain rate.

Table 2. Conversion table of vane rotation speed and shear strain rate

Vane rotation speed	n (rpm)	1/3	1/4	1/6	1/8	1/12	1/24
Shear strain rate	$\dot{\gamma}\left(s^{-1}\right)$	0.072	0.054	0.036	0.027	0.018	0.009

2.3 Results

The "Tait equation" (Dymond and Malhotra 1988) is now widely used to establishes a nonlinear relationship between density of compressible liquid and pressure under isothermal conditions. Under undrained condition, the relationship between muck density and environmental pressure was established by simplified Tait equations denoted as Eqs. (3)–(4) (ANSYS 2017).

$$\left(\frac{\rho}{\rho_0}\right)^n = \frac{K}{K_0} \tag{3}$$

$$K = K_0 + n\Delta p,\ \Delta p = p - p_0 \tag{4}$$

where p_0 is the reference pressure defined as atmospheric pressure, ρ_0 is the reference density defined as muck density at atmospheric pressure, K_0 is the reference bulk modulus, n is the density index. Plug data points in Fig. 3(a) into Eq. (3) and (4), the above coefficients are computed as $\rho_0 = 2027.1$ kg/m^3, $K_0 = 8.48 \times 10^2$ kPa, $n = 96.26$.

Vane shear tests in variable shear strain rates revealed that the residual shear stress of foam-conditioned gravelly sand increases linearly with increasing shear strain rate like Bingham model, which is written as follows:

$$\begin{aligned} |\tau| \le \tau_y,\ \dot{\gamma} = 0 \\ |\tau| \le \tau_y,\ \tau = \tau_y + \mu\dot{\gamma} \end{aligned} \tag{5}$$

where τ_y is the yield stress, μ is plastic viscosity. However, the rheological parameters of foam-conditioned gravelly sand are not constant (see Fig. 3(b)) but depends on pressure, which is denoted as Eq. (6)–(7).

$$\tau_y(p) = \tau_{y0} + a \cdot p^b \tag{6}$$

$$\mu(p) = \mu_0 + c \cdot p \tag{7}$$

where τ_{y0}, μ_0 are rheological parameters at atmospheric pressure, a, b, c are corresponding coefficients. Through least square method it can be fitted that $\tau_{y0} = 0.74$ kPa, $a = 1.34 \times 10^{-3}$, $b = 1.12$, $\mu_0 = 8.29$ kPa·s, $c = 4.48 \times 10^{-2}$.

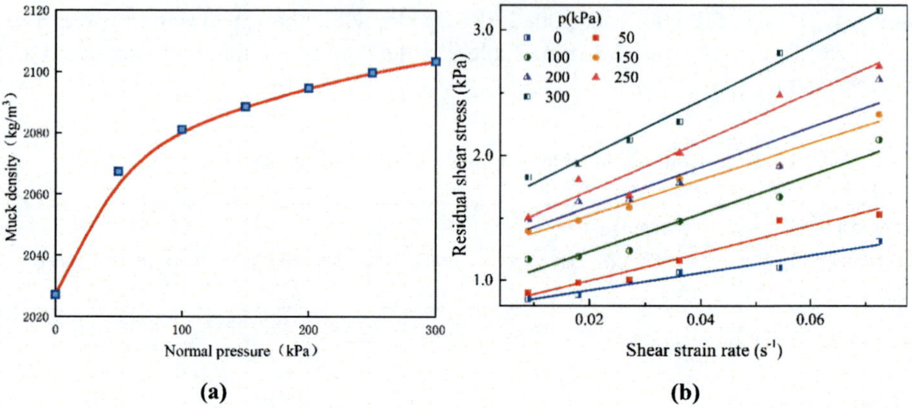

Fig. 3. Properties of foam-conditioned gravelly sand as a compressible modified Bingham fluid: (a) Density with varying normal pressure; (b) Variation in the residual shear stress with shear strain rate under different normal pressure

3 Modelling of Full Scale EPB Shield Machine

3.1 Geometric Model

In this paper, geometric model consisting of cutterhead with rotators, excavation chamber and screw conveyor was established according to EPB shield (see Fig. 4a). Table 3 shows the sizes of components, which are consistent with the EPB shield used in the tunnel project of Kunming Rail Transit. To discharge muck from the chamber, the screw extends to the chamber 0.5 m.

Table 3. Geometry size of components of EPB shield machine

Cutter head	Thickness	0.58m	**Screw conveyor**	Pitch of screw	0.65m
	Opening ratio	34%		Thickness of blade	0.08m
excavation **Chamber**	Diameter of cross section	6.44m		Diameter of blade	0.80m
	Depth in the center zone	0.82m		Diameter of rod	0.20m
	Depth in the peripheral zone	0.96m		Inner diameter of conveyor cylinder	0.80m
				Length of conveyor cylinder	11.00m
				Inclination of conveyor cylinder	21°

During excavating and discharging, it is assumed that muck is continuously and homogeneously filled in the chamber and screw conveyor. Therefore, the inside area of the geometric model mentioned above was set as computational fluid domain (see Fig. 4b). In addition, the disturbed domain ahead of cutterhead was included in the computational domain, which was supposed to be a cylinder in the CFD model.

In the EPB shield, muck moves following the rotating components, such as cutterhead, rotators and screw. Moving mesh method was used to simulate the effect of components on muck movement (ANSYS 2017). In detail, the computational muck domain was divided into several partitions. The partitions near the rotating components were moving zones, and the rest were stationary zones. It is critical to define a pair of interfaces between adjacent zones to realize the continuous flow of muck. Totally 1.2 million meshes were generated in mesh discretization.

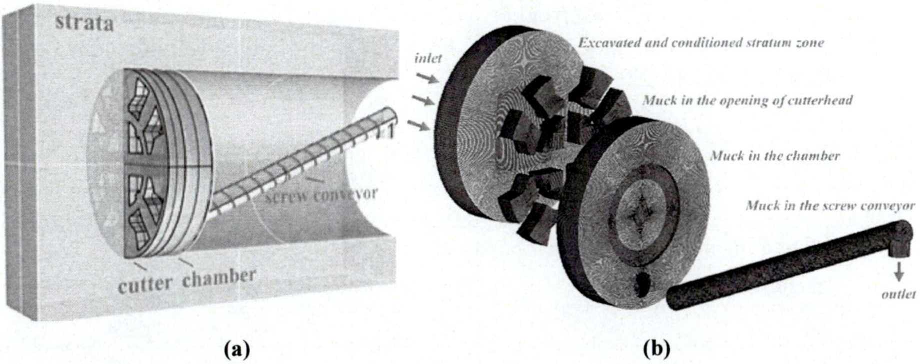

Fig. 4. Model and meshing of EPB shield machine: (a) Geometric model; (b) Partitions of computational meshing zone

3.2 Boundary Conditions

The inlet of the CFD model located in the disturbed domain was defined as total pressure boundary. The total pressure at inlet was the sum of static pressure generated by weight of overlying strata and dynamic pressure generated by muck flow and it was supposed to be linearly distributed along the depth. In model initialization, the total pressure was firstly set as static lateral stratum pressure and was revised by actual monitoring pressure at bulkhead.

The outlet of the CFD model located at the exit of screw conveyor was set as static pressure boundary, whose value is equal to atmospheric pressure. When tunnelling in test section, the average rotation speed of cutterhead was 1.4 rpm and that of screw was 7.0 rpm. Therefore, the wall boundaries of the corresponding components in the CFD model were given the same rotation speed.

Besides, pressure dependent density and rheological parameters of foam-conditioned gravelly sand obtained from laboratory tests (see Sect. 2.3) was used as fluid properties in CFD modelling.

Calculations are conducted using ANSYS-FLUENT CFD. Without consideration of muck thermodynamic properties., the governing equations are mass and momentum conservation equation, which are solved by the finite volume method. The time step size is 1×10^{-4} s.

3.3 Validation

The initial total pressure in the center of inlet was calculated as 3.03×10^2 kPa. The calibration method is that revising the total pressure at the center of inlet continuously to make the simulated time average static pressure of monitoring points close to the actual value. After calibration, the errors between measured and simulated monitoring pressure were all less than 5% with the total pressure at the center of inlet being defined as 2.83×10^2 kPa, as shown in Table 4. Furthermore, the average actual tunnelling speed in test section is 33.17 mm/min but the average speed within 120 s of simulated tunnelling is 31.53 mm/min. The error between measured and simulated tunnelling speed is only 4.94%. It can be concluded that the CFD model is available to simulate the muck movement in EPB shield.

Table 4. Comparison of in-situ test value and simulation value of monitoring muck pressure near bulkhead

	Monitoring muck pressure near bulkhead		
	In-situ value ($\times 10^2$ kPa)	Simulation value ($\times 10^2$ kPa)	Relative error
MP1	2.498	2.385	4.52%
MP2	2.719	2.599	4.41%
MP3	2.776	2.785	0.32%
MP4	2.904	3.015	3.82%
MP5	3.034	3.154	3.95%

3.4 Muck Spatial Distribution Characteristics Inside Chamber

As is shown in Fig. 5(a), each isosurface of muck static pressure is horizontal during downtime, but the static pressure isosurfaces near rotators and screw fluctuates due to rotating components during tunnelling (see Fig. 5b). This agrees with observations made in Dang and Meschke (2020).

Refer to horizontal auxiliary line, the isosurfaces of static pressure in the chamber inclines to the left in tunnelling simulation (see Fig. 5b), indicating that static pressure on the right side of chamber is higher than the left side. The field test and DEM simulation of tunnelling have also found the pressure imbalance between the left and right sides of the chamber owing to the compressible muck movement following with rotating cutterhead and rotators (Bezuijen and Talmon 2014; Qu et al. 2019).

In terms of muck movement in the chamber, Fig. 6 shows muck flow characteristics at different cross sections in the chamber. Muck mainly flows with the rotating cutterhead at section I-I near cutterhead while muck flows with the rotating rotators at section II-II. Muck clockwise moves from top to bottom of the chamber and then outflows from the screw conveyor inlet. The flow path of muck in the chamber leads to the compression of muck on the right side and the loose of muck on the left side, thus imbalance pressure will happen inside chamber during tunnelling as is mentioned above.

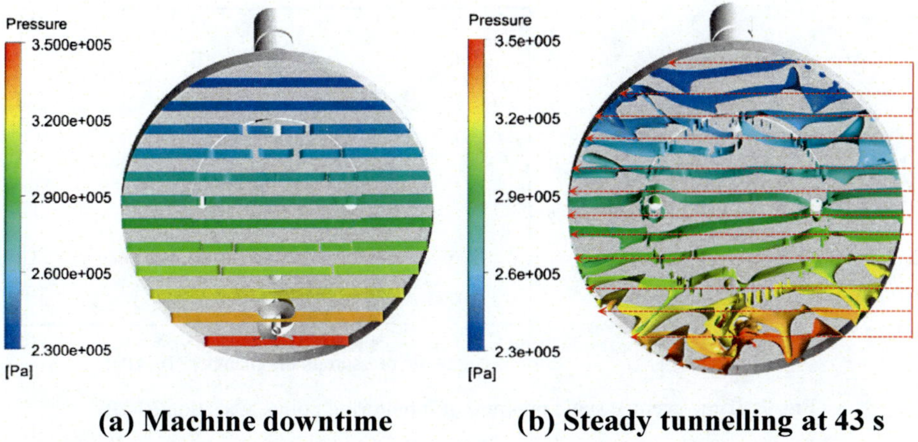

Fig. 5. Static pressure isosurface distribution in the chamber at different instants

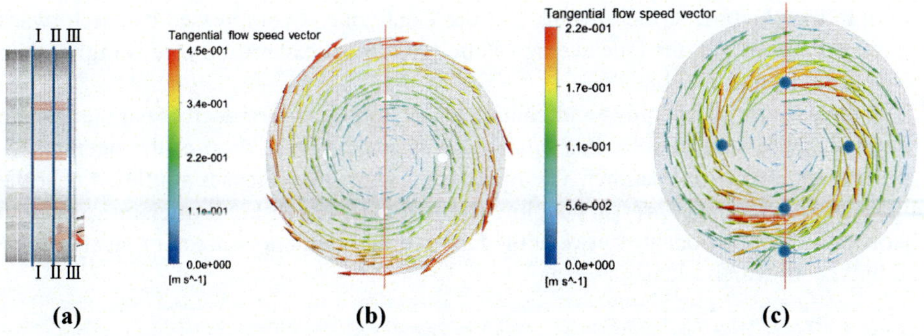

Fig. 6. Muck flow speed distribution in the chamber: (a) Monitoring sections; (b) Tangential speed vector distribution at section I-I; (c) Tangential speed vector distribution at section II-II

Furthermore, in order to eliminate the disturbance of openings on static pressure distribution, select 15 monitoring points arranging circlewise at section III-III in CFD model (Bezuijen and Talmon 2014) and compare the time average simulated static pressure of each monitoring point within 120 s to evaluate the pressure imbalance characteristic in the chamber. Figure 7 reveals that the static pressure difference on the left and right side increases with the increasing depth of the chamber. It is worth mentioning that the pressure imbalance on the left and right sides in the chamber, particularly in the bottom of the chamber, may destroy the excavation face stability, so the adverse effect of the imbalance should be considered in the adjustment of cutterhead rotation speed and muck compressibility.

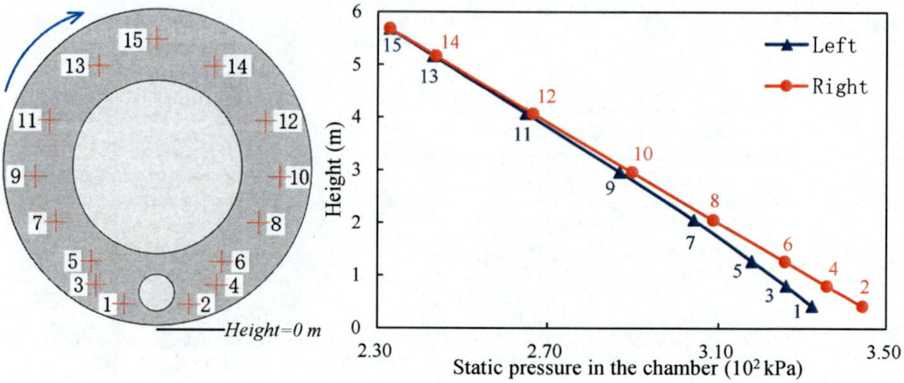

Fig. 7. Time average static pressure of monitoring points at section III-III

Muck pressure profile along the shield at steady tunnelling at 43 s is plotted in Fig. 8. When the cutterhead moves forward during tunnelling, the cutterhead panel blocks muck inflow to the chamber, which means that the front zone is compressed by the forward cutterhead. Therefore, the face average static pressure at excavation face is higher than that in front of the shield.

In the cutterhead openings and chamber, the static pressure decreases gradually to make muck inflow available. Particularly, the static pressure drop at the openings is rapider than that inside chamber. The muck static pressure increases a little bit near the bulkhead. Moreover, the pressure decline coefficient (Meng 2012) is often introduced to characterize the relationship between the face average static pressure at excavation face and bulkhead, which is calculated as

$$\alpha = \frac{\overline{p}_b}{\overline{p}_e} \tag{8}$$

where α is the pressure decline coefficient, $\overline{p_b}$ is the face average static pressure at bulkhead, $\overline{p_e}$ is the face average static pressure at excavation face. The pressure decline coefficient is related to tunnelling parameters, cutterhead opening pattern and ratio, muck flow behaviour, and so on. Meng (2012), Qu et al. (2019) pointed out that the pressure decline coefficient is generally between 0.7 and 0.8 using laboratory scale model or DEM model. However, the pressure decline coefficient is 0.99 using CFD model, which is similar to the investigation from CFD simulation by Dang and Meschke (2020). The assumption of the chamber and screw conveyor evenly filled with muck that is different from the actual situation probably leads to coefficient deviation.

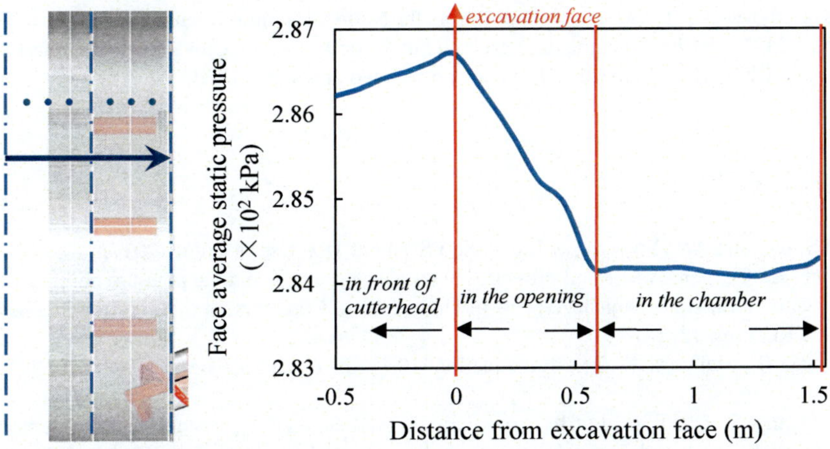

Fig. 8. Average static pressure distribution when muck flows inside the chamber

4 Conclusions

(1) Foam-conditioned gravelly sand was regarded as compressible modified Bingham model. Pressure dependent density, yield stress and plastic viscosity and their coefficient were obtained from vane shear tests and lateral compression tests.

(2) The CFD model of muck inflow and outflow in EPB shield used in Kunming Rail Transit was established. The total pressure of model inlet was revised according to the geological data and the field-measured monitoring pressure at bulkhead sensors. When the total pressure at the center of inlet boundary is 2.83×10^2 kPa, the errors between the field-measured and simulated static pressure at bulkhead are all less than 5% and the error between the field-measured and simulated average tunneling speed in test section is only 3.86%, which proves that the CFD model is feasible to reflect the dynamic tunnelling process.

(3) The CFD simulation presents the muck static pressure and flow speed distribution inside shield, which cannot be observed on site. From the disturbed domain in front of the excavation face to the bulkhead, the static pressure increases first and then decreases, but the static pressure decline coefficient is close to 1. The static pressure distribution in the shield is disturbed by the rotating shield components. The static pressure imbalance between the left and right sides occurs inside chamber, particularly at the bottom of chamber. The muck flows around the chamber axis with rotating components until it is discharged by rotating screw, which leads to the pressure imbalance.

This paper reveals that the CFD model can simulate the complex dynamic process of EPB shield tunneling to present the spatial distribution of muck static pressure and flow speed as well as their interaction.

Acknowledgments. The financial support from the National Natural Science Foundation of China (No. 52022112) and the Fundamental Research Funds for the Central Universities of Central South University, China (No. 2020zzts152) are acknowledged and appreciated.

References

ANSYS, Inc. and ANSYS Europe, Ltd. ANSYS Fluent 18.0 User's Guide (2017)

Bezuijen, A., Talmon, A.M.: Soil pressures at the cutting wheel and the pressure bulkhead of an EPB-shield. In: Geotechnical Aspects of Underground Construction in Soft Ground, pp. 523–529. CRC Press (2014)

Dymond, J.H., Malhotra, R.: The Tait equation: 100 years on. Int. J. Thermophys. **9**(6), 941–951 (1988)

Hu, Q., Wang, S., Qu, T., et al.: Effect of hydraulic gradient on the permeability characteristics of foam-conditioned sand for mechanized tunnelling. Tunnel. Undergr. Space Technol. **99**, 103377 (2020)

Hu, W., Rostami, J.: A new method to quantify rheology of conditioned soil for application in EPB TBM tunneling. Tunnel. Undergr. Space Technol. **96**, 103192 (2020)

Lin, L.: Research on opening characteristic and cutter's layout method of EPB shield cutterhead. Central South University (2013)

Liu, P., Wang, S., Hu, Q., et al.: Effect of dispersant on the tangential adhesion strength between clay and metal for EPB shield tunnelling. Tunnel. Undergr. Space Technol. **95**, 103144 (2020)

Qinglin, M., Shouju, L.: Experimental investigation on viscoplastic parameters of conditioned sands in earth pressure balance shield tunneling. J. Mech. Sci. Technol. **25**(9), 2259–2266 (2011)

Meng, Q.: Research and application in control of conveying mechanism of screw conveyor under pressure of EPB shield. Dalian University of Technology (2012)

Qu, T., Wang, S., Hu, Q.: Coupled discrete element-finite difference method for analysing effects of cohesionless soil conditioning on tunneling behaviour of EPB shield. KSCE J. Civ. Eng. **23**(10), 4538–4552 (2019)

Dang, T.S., Guenther, M.: A shear-slip mesh update – immersed boundary finite element model for computational simulations of material transport in EPB tunnel boring machines. Finite Elements Anal. Design **142**, 1–16 (2018)

Dang, T.S., Guenther, M.: Influence of muck properties and chamber design on pressure distribution in EPB pressure chambers - Insights from computational flow simulations. Tunnel. Undergr. Space Technol. **99**, 103333 (2020)

Wang, S., Liu, P., Hu, Q., et al.: State-of-the-art on theories and technologies of soil conditioning for shield tunnelling. China J. Highway Transp. **73**, 244–251 (2020)

Yi, Y., Li, X., Su, W.: Experimental investigation on rheological behaviours of bentonite- and CMC-conditioned sands. KSCE J. Civ. Eng. **24**(6), 1914–1923 (2020)

Estimating the Damage of Earthquake Using RADIUS Model (Case Study: Tehran)

Hossein Etemadfard[1](✉) and Maryam Moradi[2]

[1] Civil Engineering Department, Faculty of Engineering, Ferdowsi University of Mashhad (FUM), Mashhad, Iran
etemadfard@um.ac.ir
[2] K.N.Toosi, University of Technology, Tehran, Iran

Abstract. By its specific geographical and climate conditions, Iran benefits from high environmental and climate variability. Besides abundant relative advantages, this variety has increased the potential of different natural risks. Earthquake, as one of the most significant geotechnical risks, threatens most regions throughout Iranian Plateau. The importance of earthquake risk in Iran has been increased by the increased development of cities and the concentration of population. Tehran, as the most important city based on its population, economic and social infrastructures, and because it has been surrounded by several active faults as well as located on some active ones, exposed to high risk against earthquake risk. In this case, one of the requirements of urban management in Tehran is studying the seismic risks of this town as well as estimating its probable damages. For this reason, region 3 of district 1 of Tehran municipality having many old textures than other districts have been chosen as a region of study. The methodology of this study, data collection, and analysis are conducted by methods based on database benefiting from RADIUS Program, and there have been considered two scenarios based on both faults in the North Tehran as an origin for estimating the risks from the probable earthquake in the region. Results indicate that maximum risk might be considered in the scenario of fault located in North Tehran.

Keywords: RADIUS model · Earthquake · Damage · District 1 of Tehran municipality · Scenario

1 Introduction

Natural disasters like earthquakes, floods, and drought are among the cases that continuously endangers people throughout the world. Various values of such risks depending on factors such as location, physical conditions of living place, and human behaviors in its living place. Public awareness of risks, knowing it, understanding its factors as well as any information related to factors endangering an individual or community than others, can help to find the alternatives for reducing such risks. Possessing such estimations from natural risks threatening a human community is considered a powerful tool for urban designers and or natural sources managers in high-risk regions.

B. Fatahi et al. (Eds.): GeoChina 2021, SUCI, pp. 49–62, 2021.
https://doi.org/10.1007/978-3-030-79854-3_4

The occurrence of a severe earthquake caused the human to think to prepare an infrastructure plan for reducing its risks. National tectonic features have discussed the earthquake as one of the most dangerous factors destructing human life. Historical studies indicate that most regions throughout our country have suffered from this natural disaster financially and life-threatened. According to the UN report, in 2003, among the world's countries, Iran had one of the first ranks in the field of several earthquakes with an intensity of 5.5 and earthquake vulnerability and some kills due to this disaster. According to the same report, in Iran, the earthquake is dominated among other natural disasters (UNDP 2004).

What turns the earthquake into a disaster is the lack of knowledge by man and its disability for encountering and combating with. This issue is generally strengthened by the widest reckless interference of humans in natural environments such as indiscriminate construction of faults, lack, and or ignorance of construction standards.

As one of the most populated cities in the country, Tehran has been located on some main faults including Nasga-Fasham, North Tehran, Niavaran, Down Telo, Mahmoudieh, Shian and Kothar, North Rey, South Rey, Kahrizak, Garmsar, Pishva, and Parchin faults; there are however many other side faults through Tehran such as Narmak, Shadabad, Davoudieh, Abbasabad, Baghe Feiz. According to the high size of faults under Tehran and the historical background for activities of such faults, it can be said that one day not long ago, Tehran will face a massive earthquake. In this case, based on the presence of many faults in the neighborhood and region 3 of District 1 of Tehran's Municipality, the vulnerability of these regions will be increased against earthquake risk. Another case that amplified the vulnerability of urban contexts in this region is the presence of self-organizing and old textures in this region of Tehran with lower stability against earthquakes as the main characteristics of such urban textures. Another issue of such textures is unsuitable and limited access to them complicating the relief of its inhabitants after the earthquake and can trigger human tragedy under critical situations. Some of such old textures include Darband, Baghe Shater, Abak, and Imamzadeh Gasim. As mentioned above, it is strictly necessary to evaluate and analyze the risks of a probable earthquake in this region of Tehran by making a proper model and using types of spatial and non-spatial data as well as related analyses to be prepared against such natural disaster and manage the crises resulted from natural disasters during a systematic process.

So far, there have been conducted many analyses and assessments for estimating the damages with evaluating the vulnerability against earthquake in terms of various studies; each of which is used of different models and methods based on objectives and theories.

Tavakoli and Tavakoli (1993) took one of the most important actions for determining the physical vulnerability of constructions in Iran and its results came to the assessment of failure curves for three different types of buildings based on Roodbar and Manjil earthquakes. They studied the damage to villages close to the center of the earthquake of 1990 in Manjil and extracted the relation between maximum land acceleration and damages of buildings.

In a study, benefiting from two models RISK_UE and TOPSIS Fuzzy and GIS as well as spatial and non-spatial measures, Amini (2010) studied the vulnerability of residential places of district 9 of Tehran against the earthquake as well as investigating the vulnerability of this region under different seismic intensities and concluded that this region is vulnerable against a probable earthquake.

Another study for estimating the vulnerability is the Japan International Cooperation Agency (JICA) project for Tehran. During this project, the vulnerability of metropolitan Tehran has been investigated for different physical and human cases as well as for specific places based on failure curves prepared by Tavakoliha (JICA 2001).

Using the Analytical method of Assessing Seismic Vulnerability using AHP and GIS, Hataminejad et al. (2009) investigated the vulnerability of district 10 of Tehran; for this reason, he used indices such as type of materials, lifespan of structure, density and communication grid.

In a study, Giovinazi et al. (2006) initially investigated different vulnerability models such as RISK_UK and different scenarios of vulnerability and then investigated the vulnerability of Linguria in Italy using RISK_UK and provided vulnerability scenarios.

In a study, Gulati (2006) while comparing the models for vulnerability assessment of the city against earthquake- RADIUS (Villacis and Cardona 1999) & TELES-evaluated the risk of buildings constructed on Dehradon town, India against the earthquake using Hazus model and conclude that due to its various data and variables used there, Hazus model can be used as a suitable model for evaluating and reducing the vulnerability of towns against the earthquake.

In another study, Lantada et al. (2009) modeling the vulnerability of Barcelona town, using the RISK_UE model, and applying existing models in the field of estimating the earthquake damages, evaluating the human and economic damages in this town.

Tang and Wen (2009) in their study used Artificial Neural Grid for evaluating the earthquake risk in Diang, China-based on the development of GIS and Artificial Grid. This system is applicable for diagnosing the seismic weakness of structures under the pre-earthquake conditions, fast evaluation of earthquake damages, and providing intelligent instant conditions for a public response during and after the earthquake.

As a powerful and flexible framework to characterize the temporal seismicity changes and reveal unexpected patterns, Poisson's hidden Markov model provides a better understanding of the nature of earthquakes and is studied recently. Poisson hidden Markov model is used to predict the earthquake hazard in Bilecik (NW Turkey) because of its important geographic location. The annual frequencies of earthquakes occurred within a radius of 100 km area centered on Bilecik, from January 1900 to December 2012, with magnitudes (M) at least 4.0 are modeled by using Poisson-HMM. The hazards for the next 35 years from 2013 to 2047 around the area are obtained from the model by forecasting the annual frequencies of $M > 4$ earthquakes (Eda Can et al. 2014).

The probability obtained by the Bayesian Networks was used to determine the disaster chain probability and hazard intensity of the earthquake events, while ArcGIS was used to produce the disaster chain hazard map in a study done by Han et.al. The performance of the Bayesian Network model was measured by error rate and scoring rules. The confirmation of the outcomes of the disaster chain hazard assessment model shows that the model demonstrated good predictive performance based on the area under the curve, which was 0.7929. From visual inspection of the produced earthquake disaster chain hazard map, highly hazardous zones are located within a 15 km radius from the Tianchi center, while the northern and the western parts of the studied area are characterized mainly by "very low" to "low" hazard values (Han et al. 2019).

There is a review done of artificial neural network (ANN) applications for earthquake prediction (77 articles, 1994–2019 period) and two emerging trends found: an increasing interest in this domain, and a complexification of ANN models over time, towards deep learning. Despite apparent positive results observed in this corpus, they demonstrate that simpler models seem to offer similar predictive powers, if not better ones. Due to the structured, tabulated nature of earthquake catalogs, and the limited number of features, simpler and more transparent machine learning models seem preferable at the present stage of research. Those baseline models follow first physical principles and are consistent with the known empirical laws of Statistical Seismology, which have minimal abilities to predict large earthquakes (Mignan and Broccardo 2019).

Observing the studies conducted in this field, one can state that there are many models developed for evaluating and reducing the vulnerability of towns and cities against earthquakes. Using a suitable model and algorithm by which one can deal with the assessment of city's vulnerability against earthquake has many advantages and great importance and besides predicting the damages and casualties, it may provide the conditions to take necessary action for reducing the vulnerability against earthquake with highly important for emergency reactions and crisis management.

2 Case Study

For this study, region 3 of district 1 of Tehran's municipality with a high size of the old texture than other regions through the district was selected that has been located in the Northern part of metropolitan Tehran and the southern foothill of the Central Alborz Mountains. Municipality of District 1 has 10 regions and 26 neighborhoods; its population, according to the census of 2011 is 379962 people and the population of region 3 is about 59673. Figure 1 indicates the situation of the region of study together with other ten regions of district 1 of Tehran Municipality.

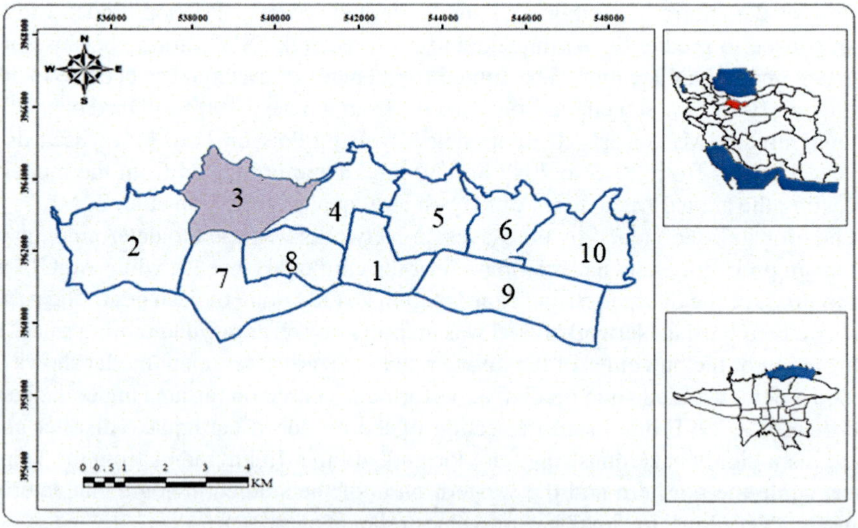

Fig. 1. Situation of the region of study

Based on the required input for model and available data, it includes land-use of buildings (https://map.tehran.ir), number of floors, the total population of the region and vital arteries (street grid, water reservoir, power and telecommunication posts, water and sewage pipelines, and fuel stations). In addition, ArcGIS and RADIUS software have been used in this study.

3 RADIUS Model

RADIUS approach was used in 1996 aiming to prepare earthquake scenarios and preparing a plan for cities vulnerable to earthquakes in the developing countries. The main objective of the RADIUS project, supported by the UN, is promoting public awareness and making a scientific and applied tool for reducing the earthquake risk in the urban regions. By some modifications, this approach has been used as software for estimating the damage and preparing the seismic scenario. This model is applicable for information sharing plan for all beneficiaries in the city. RADIUS plan comes with the following objectives (Villacis and Cardona 1999):

1. Designing a tool for Seismic Risk Management being able to respond to the earthquake-prone cities;
2. Directing the comparative studies for perceiving the understanding of seismic risk in the urban regions through the world;
3. Data sharing for reducing the seismic risk effects in the urban regions;
4. Preparing a plan for promoting the urban structure such as reinforcing thc buildings and vulnerable infrastructures, making the open spaces and emergency roads safe;
5. Preparing the rescue facilities, firefighting, and emergency transportation;

RADIUS is a program working under EXCEL environment and the user must input size, boundaries of the region of study by gridding, the total population of the region of study, total buildings, and type of structures used in the buildings, type of soil of the studied region, information about vital arteries of the region under study, selecting the seismic scenario and its parameters in this software. Then this program will review the validity of input data and analyze it. The outputs of this program include:

1. Seismic intensity by PGA and MMI intensity;
2. Damages made to buildings;
3. Damages made to vital arteries;
4. Casualties, including the number of kills and injuries;
5. Tables and maps showing the results by subject.

One of the main objectives of this project was the development of an experimental tool for risk management in the city. RADIUS model for destruction of buildings can be separated into 10 stages including determination of seismic scenario; calculating the casualties by using a function; calculating the strengthening resulted from local conditions of the soil using soil map; turning the PGA to Modified Mergali Intensity; applying vulnerability function for types of buildings, using vulnerability unction for types of infrastructures, using vulnerability function for casualties; applying buildings costs data and combining to vulnerability for calculating the destructions for different return periods; combining the destruction data for different return periods and calculating the risks by adding the destructions from these periods; combining the data and summarization (Villacis and Cardona 1999). Figure 2 indicates the general process of estimating the damage in this program.

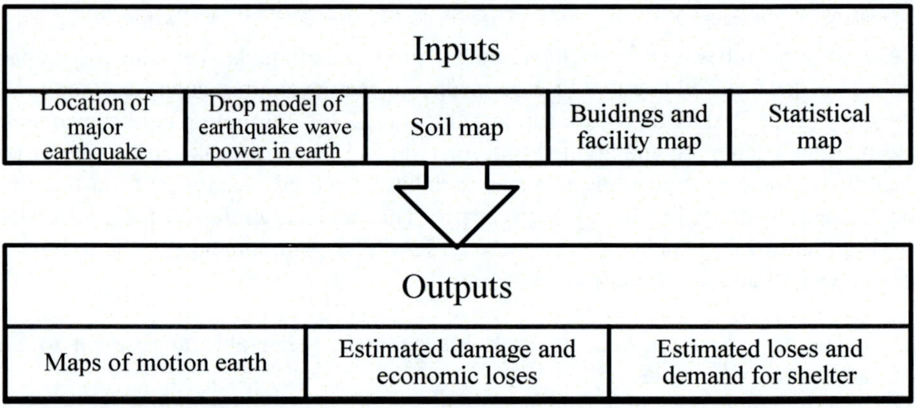

Fig. 2. Process of damage estimation by RADIUS Program (Sarvar et al. 2011)

4 Implementation

Stages for estimating the damages from the earthquake in District 1 of Tehran Municipality by RADIUS Software include six main stages detailed as below (Fig. 3 depicts the whole process).

4.1 Regional Gridding

Because in civil regions, any part of the region come with different properties such as the situation of ground, type and different usage of buildings as well as different statistical data for estimating the damage, region 3 was divided into 15 equal grids with plots of 500 × 500 m and data required by program inputted in the software separated by any grid. The region was gridded in ArcGIS software and based on the boundaries of the district and distributing the existing usages. Figure 4 indicates the region grids in RADIUS Model.

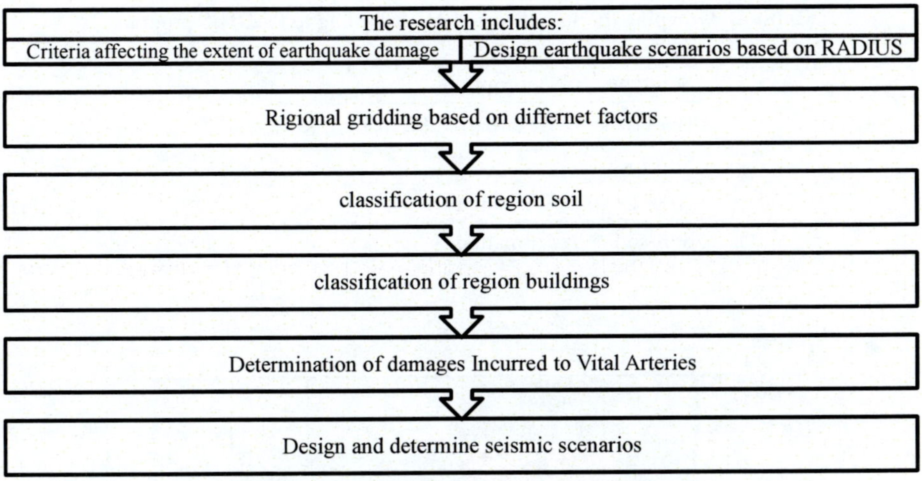

Fig. 3. The flowchart of designed framework for this study

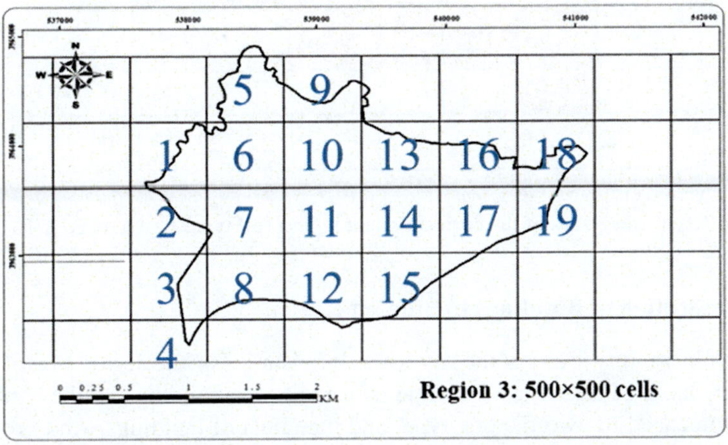

Fig. 4. Gridding the region according to RADIUS Model (from 1 to 19 cells)

4.2 Determination of Soil Situation of the Region

The situation of regional soil is effective in the damage from earthquakes because the ground directly amplifies the ground vibration and its seismic effects. RADIUS software proposes a simple classification for soil type and in such classification soil type is grouped into 4 classes including hard rock, soft rock, average soil, and soft soil and aggravating factors of seismic intensity has been also classified by related software; Table 1 indicates the grouping of the regional soil. Figure 5 shows the results of grouping the soil of region.

Table 1. Grouping the soil of region (3) according to RADIUS program

Soil type	Description
Hard rock	Hezardarreh Formation, Rock bed
Soft rock	---
Average soil	Formation C
Soft soil	Formation B

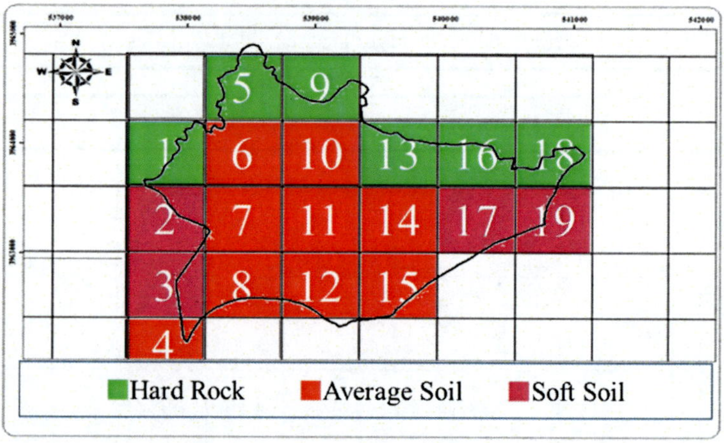

Fig. 5. Different categories of the correspondent region soil types based on RADIUS program

4.3 Classification of Buildings in Studied Region

Building vulnerability is one of the key issues in seismic risk; the destruction of buildings is the main cause of casualties. Damage incurred to the buildings by the earthquake is directly influenced by two factors, type, and materials of buildings. And grouping the buildings is the first step for determining their damages. RADIUS software classifies the buildings based on the type of materials used in construction, usage, and the number of floors. Knowing the number of existing buildings in any grid is necessary for calculating the rate of damages and different types of existing buildings in any grid were calculated by ArcGIS and inputted into RADIUS Software. Table 2 and Fig. 6 indicate the classification of buildings according to the RADIUS model as well as their situation in the studied region.

Table 2. Grouping the buildings according to RADIUS Program

Building	Description	No.	Percentage
Residential	Type 1: buildings with masonry and materials non-resistant against probable earthquakes, made out of brick, thatch and ceilings and walls without tie beam	15	0.27
	Type 2: hybrid structures, non-standard structures and non-compatible with construction guidelines, height up to 3 floors	3178	57.6
	Type 3: Buildings with steel or concrete skeleton, incompatible with construction guidelines, height up to 4 to 6 floors	1307	23.7
	Type 4: Buildings with steel or reinforced concrete skeleton, engineered, new buildings with several floors	741	13.43
Training	Type 1: construction of schools up to 2 floors	19	0.34
	Type 2: construction of schools more than 2 floors	12	0.3
Medical	Type 1: small and average hospitals	8	0.14
	Type 2: Big hospitals	1	0.02
Commercial	Purchase centers	233	4.2
Industrial	Industrial facilities and buildings	0	0

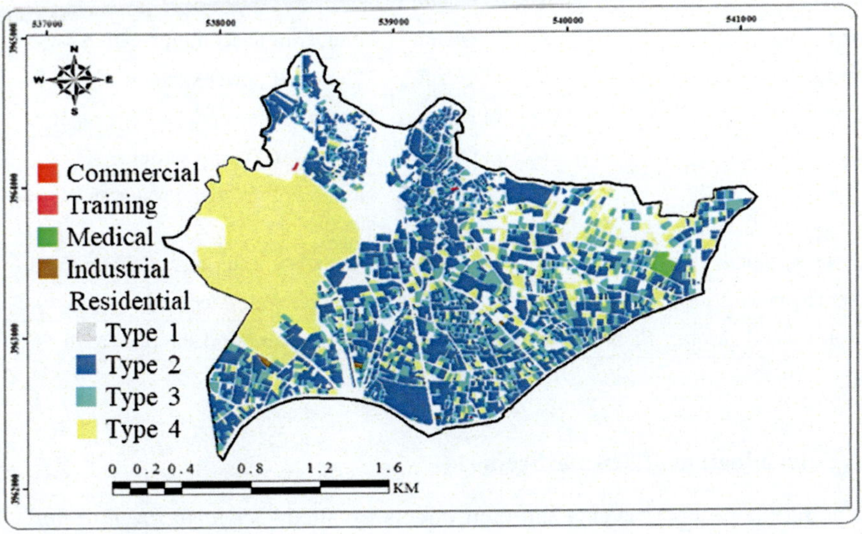

Fig. 6. Grouping the buildings according to RADIUS program

4.4 Determination of Damages Incurred to Vital Arteries

Should vital arteries such as watering networks, electricity networks or other communicating networks damage due to earthquake; it follows with next damages such as damages to street networks and streets are particularly important as one of the main urban elements, immediately after the earthquake because it is necessary to evacuate the injured persons. Such evacuations must be conducted by inter-urban roads, intra-urban streets, and side streets, and upon closing one of the main roads and or even sideways, the damages from the earthquake will be multiplied and it is possible to take longer time returning to the normal situations. To estimate the damages from vital arteries, the RADIUS program uses general statistics and total vital arteries through the region and its relation with seismic intensity; i.e. the more the seismic intensity is, the more is the damage incurred to vital arteries for each facility in terms of a curve. Table 3 indicate the grouping of vital arteries according to the RADIUS model in which its situation in the region under study has been also indicated.

Table 3. Situation of vital arteries in region 3, District 1 of Tehran municipality

Vital arteries	Rate	Unit	Description
Road of type 1	8	Km	Local roads surrounding the regions
Road of type 2	2	Km	Highways and roads
Bridge	0	Quant	-
Tunnel	0	Quant	-
Power 1	2	Quant	Number of telecommunication towers
Power 2	25	Quant	Number of power stations
Water 1	12	Km	Length of water and sewage pipeline
Water 2	3	Quant	No. of water pump stations
Water 3	0	Quant	Number of industrial water and sewage stations
Water reservoir 1	4	Quant	Number of water reservoirs and seals
Water reservoir 2	0	Quant	Number of elevated tanks
Petroleum and gasoline fuels	2	Quant	The number of oil stations etc

4.5 Determination of Seismic Scenario

Predicting the probable events and their effects is called "*Scenario Preparation*". The scenario is mostly prepared for making the plan and preparation for disasters in the disaster-prone communities and helping its perception for the future. Earthquake scenario means the determination of its magnitude, intensity, and other seismic parameters considered a probable earthquake in the region by software. Parameters input to the system for determining the earthquake scenario include the spatial situation of earthquake, depth of the earthquake, its magnitude and time of occurrence; because casualties and damages are different depending on its occurrence at night or day and what damages incurred to public buildings, schools and organizations. In this study, or making a proper image of occurrence of damages in this region and preparing types of probable damages due to earthquake in this region, there has been considered two earthquakes resulted from the activity of two faults, Mosha and North Tehran; then, there was developed two models for scenario earthquakes by RADIUS Program as indicated in Table 4, and time of earthquake considered as 1:00 am. It must be noted that by the occurrence of the earthquake at night, the casualties will be increased; however, the number of buildings destructed are the same day and night.

Table 4. Specifications of models of earthquake scenario

Specifications	Mosha fault model	North Tehran fault model
Distance to the studied region	Eastern North	North
Earthquake magnitude	7.2	7.2
Earthquake depth	5 km	5 km
Distance from reference grid (7)	20 km	0.9 km

5 Results and Discussion

After implementation of model and producing the output, the results have been shown in the Table 5. It indicates the results of damages and casualties from the occurrence of earthquakes according to scenarios considered in the studied region. Figures and statistics of this table indicate that according to the scenario of Mosha fault; there will be destructed about 1009 buildings, with 157 casualties and 2111 injured.

Table 5. Damage and casualty rate incurred by region 3 under different scenarios

Earthquake scenario	Damages and casualties	Region 3
Mosha fault model	Building	1009
	Kills	157
	Injuries	2111
North Tehran fault model	Building	2508
	Kills	908
	Injuries	7175

According to the scenario of North Tehran's fault, there will be also destructed about 2508 buildings, with 908 casualties and 7157 injured persons. To have a spatial view related to the casualties for the studied region with determining the figures and statistics in the table spatially, Figs. 7 and 8) indicate the percentage of destructed buildings for related scenarios.

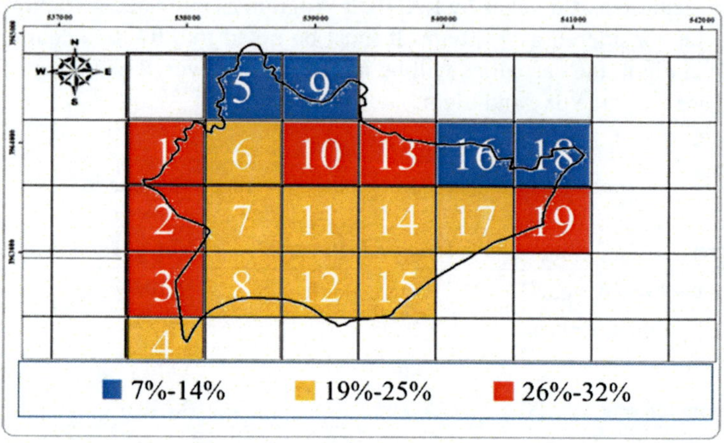

Fig. 7. Building's distribution percentages based on the scenario of Mosha fault

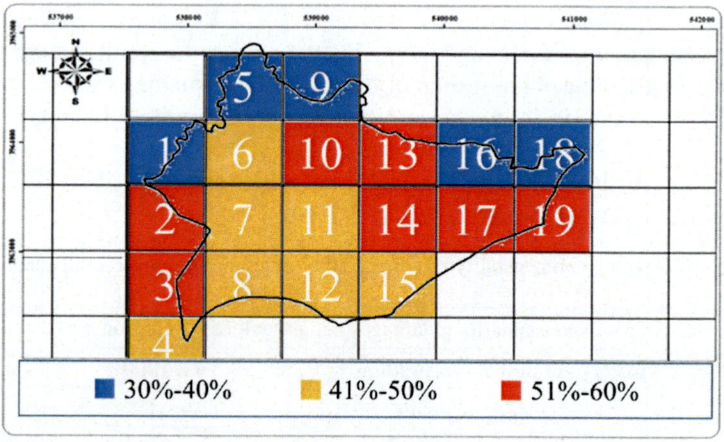

Fig. 8. Building's distribution percentages, based on the scenario of North Tehran fault

It noteworthy that the rate of casualties and injured persons are influenced by the destruction rate of buildings; for this reason, there has only been indicated the map of buildings destroyed. Table 6 indicates the results from damages in the vital arteries based on considered scenarios.

Table 6. The percentage of damages to vital arteries in the region of study, under different scenarios

Earthquake scenario	Road of type 1	Road of type 2	Water and sewage pipelines	Oil stations
Mosha fault model	0.4	0.1	2	0.3
North Tehran fault model	1.3	0.1	3.8	0.7

The statistics indicate that according to Mosha fault model, 0.5% of streets, 2% of water and sewage pipelines and 0.3% of oil stations in the region under study will be destructed and according to North Tehran fault model, 1.4% of streets, 3.8% of water and sewage pipelines, and 0.7% of oil stations will be destructed. The scenario of North Tehran fault comes with maximum damage in the region, this is because of its close distance of this region with related fault.

6 Conclusion

Tehran is the most important city in Iran and has a huge amount of population, economic and social infrastructures where surrounded by several active faults. Because of them, one of the urban management parameters is studying the seismic risks in Tehran as well as estimating its probable damages. For this purpose, region 3 in district 1 of the Tehran municipality has been selected as a pilot to research because of many old textures than other districts. The RADIUS Program has been applied on two scenarios which are based on Mosha fault or the North Tehran fault.

According to the results, the buildings will be destructed with the Mosha fault model at least 50% less than the North Tehran fault model. According to the results, the Mosha fault model will be destructed at least 40% less than the North Tehran fault model in region 3. It is also predicted that the number of deaths with the North Tehran fault model will be six times higher. Overall, the Mosha fault model has less damage in the study area.

The measurements demonstrate that concurring to Mosha fault show, 0.5% of lanes, 2% of water and sewage pipelines and 0.3% of oil stations within the region under study will be destructed and agreeing to North Tehran fault demonstrate, 1.4% of roads, 3.8% of water and sewage pipelines, and 0.7% of oil stations will be destructed. The situation of North Tehran fault comes with the most extreme damage within the locale, this can be since of its near removal of this locale with related fault.

References

Amini, J.: Analysis of vulnerability of urban residences against earthquake (a case study of district 9 of Tehran). Master thesis, remote sensing and GIS. Tarbiate Modares University, Tehran (2010)

Villacis, C.A., Cardona, C.N.: Guidelines for the Implementation of Earthquake Risk Management Projects. Geohazards International. Palo Alto, (1999)

Cardona, C.N.: Earthquake damage scenarios for thimpho Valley- implementation of case study-geo hazards International, vol. 38–42 (2002)

Eda Can, C., Gul Ergun, G., Gokceoglu, C.: Prediction of earthquake hazard by hidden Markov model (around Bilecik, NW Turkey). Cent. Euro. J. Geosci. **6**, 403–414 (2014)

Giovinazzi, S., Lagomarsino, S., Pampanin, S.: Vulnerability methods and damage scenario for seismic risk analysis as support to retrofit strategies: an European perspective. In: NZSEE Conference (2006)

Gulati, B.: Earthquake risk assessment of buildings applicability of HAZUS in Dehradun, India. Unpublished MS Thesis, ITC, the Netherlands (2006)

Han, L., Zhang, J., Zhang, Y., Ma, Q., Alu, S., Lang, Q.: Hazard assessment of earthquake disaster chains based on a Bayesian network model and ArcGIS. Int. J. Geoinf. **8**, 210 (2019)

Hataminejad, H., Fathi, H., Eshgabadi, F.: Evaluation of seismic vulnerability in the city, a case study: district 10 of Tehran . J. Res. Hum. Geograp. **68**, 1–20 (2009)

Japan International Cooperation Agency (JICA): Micro-zonation Seismic Project of Metropolitan, Tehran (2001)

Lantada, N., Pujades, L., Barbat, A.: Vulnerability index and capacity spectrum based methods for urban seismic risk evaluation a comparison. Nat Hazards **51**, 501–524 (2009)

Mignan, A., Broccardo, M.: Neural network applications in earthquake prediction (1994–2019): meta-analytic insight on their limitations. Seismol. Res. Lett. **91**(4), 2330–2342 (2019)

Sarvar, H., Amini, J., Laleh-Poor, M.: Assessment of risk caused by earthquake in region 1 of Tehran using the combination of RADIUS, TOPSIS and AHP models. J. Civil Eng. Urban. **1**(1), 39–48 (2011)

Tang, A., Wen, A.: An intelligent simulation system for earthquake disaster assessment. Comput. Geosci. **35**(5), 871–879 (2009)

Tavakoli. B., Tavakoli. A.: Estimating the vulnerability and loss functions of residential buildings. J. Int. Soc. Present. Mitig. Nat. Hazard **7**(2), 155–171 (1993)

Villacis, C.A., Cardona, C.N.: Guidelines for the Implementation of Earthquake Risk Management Projects. Geo Hazards International. Palo Alto (1999)

UNDP: Reducing Disaster Risk, a Challenge for Development, UNDP, New York (2004)

Experimental Investigation on Mechanical Properties of Cemented Expansive Soil

Zhuang Miao, Fangtong Wang, Chia Zarei, and Yong Liu(✉)

State Key Laboratory of Water Resources and Hydropower Engineering Science, Institute of Engineering Risk and Disaster Prevention, Wuhan University, Wuhan, Hubei, China
liuy203@whu.edu.cn

Abstract. Expansive soil has the characters of swell-shrink, fissured and over-consolidated, which can affect the safety of construction projects in expansive soil areas. Deep soil mixing (DSM) is a widely used ground improvement method. It can improve the mechanical properties and reduce the permeability of expansive soil effectively. In this study, a series of direct shear tests and unconfined compressive tests were conducted to study the effect of cement content and curing time on the mechanical properties of cemented soil. Relationship among the mechanical properties and cement content and curing time were analyzed qualitatively and quantitatively by the test results. Recommendations for design and construction of the DSM in expansive soil areas were also proposed in this study.

Keywords: Expansive soil · Cemented soil · Direct shear test · Unconfined compressive test · Mechanical properties

1 Introduction

Expansive soil has the characters of swell-shrink, fissured and over-consolidated, which can affect the safety of construction projects in expansive soil areas (Yu et al. 2006; Huang et al. 2009). The cracks caused by the swell-shrink behavior can lead to the deformation and damage of buildings (Zuo et al. 2007), which further affects the safety of projects (Shi et al. 2002). The total length of the Middle Route Project of South-to-North Water Diversion is about 1432 km. The length of main canal involved in the expansive soil areas is about 386.8 km, accounting for 27% of the total length (Gong et al. 2011). The existence of expansive soil affects the slope safety of the canals. Therefore, the problem of expansive soil becomes a fundamental technical problem that must be faced and solved in the South-to-North Water Transfer Project (Gong et al. 2011).

The elastic modulus, compressive strength, cohesion, and internal friction angle can increase through the cement modified method (Yu et al. 2006). The current research mostly focused on the cement modified expansive soil which the cement content is less than 10% (Khemissa and Mohanmedi 2014; Pora et al. 2017). Madhyannapu et al. (2014) improved the DSM method and applied it to the expansive soil areas. Bhadriraju et al. (2007) presented a laboratory procedure to prepare well-mixed soil binder samples for simulation of deep soil mixing to stabilize medium-stiff to stiff expansive clayey soils;

B. Fatahi et al. (Eds.): GeoChina 2021, SUCI, pp. 63–73, 2021.
https://doi.org/10.1007/978-3-030-79854-3_5

Madhyannapu (2006) proposed a stepwise procedure to simulate the field deep soil mixing process for expansive soils in the laboratory environment and then conducted a series of experiments; Madhyannapu (2006) discussed the evaluation of the effectiveness of DSM technology in mitigating shrink-and-swell behaviors of expansive subsoils under actual field conditions. In the engineering practices, Shen et al. (2007) used the deep mixing method to improve the safety of the ground; Ye et al. (2009) used the deep mixing method to consolidate deep soft soil foundation of the embankment.

Compared to the cement modified expansive soil, the deep soil can also be reinforced effectively by DSM. However, the researches on the expansive soil reinforced by DSM are limited. Therefore, a series of laboratory tests were conducted to describe the relationship among mechanical properties and cement content and curing time. The results shall be of practical value in engineering construction in expansive soil areas.

2 Laboratory Investigation

The laboratory test mainly includes sample preparation, unconfined compressive test, and direct shear test. The expansive soil used in the experiment was taken from Nanyang City, Henan Province, China (Fig. 1). The cement used is OPC 42.5 grade. The cutting ring samples with the diameter of 61.8 mm and the diameter of 20 mm used in the direct shear tests. The shear rate used herein is 0.8 mm/min by following the standard for soil test method in China (GB/T 50123-1999). Four vertical loads of 100 kPa, 200 kPa, 300 kPa, and 400 kPa was applied for the direct shear test. The cylindrical samples with the height of 100 mm and the diameter of 50 mm were used in unconfined compressive tests. The loading rate is 1%/min according to the standard for soil test method in China (GB/T 50123-1999).

The sample preparation of expansive soil is based on the standard for soil test method in China (GB/T 50123-1999). The expansive soil sample was crushed and passed through a 2 mm sieve, and placed in a drying oven. The dry density of the samples is 1.57 g/cm^3 which equal to the in-situ dry density. The weight of the expansive soil powder and deionized water was calculated, and put them in a sealed plastic bag for 12 h. The mixed soil was passed through a 2 mm sieve to reduce the influence of soil aggregates. The samples for direct shear test and unconfined compressive test were compacted in one and four layers, respectively. After that, the samples were put into a saturator for vacuum saturation.

For cemented expansive soil, the cement content is defined as the mass ratio of cement to soil; the water content is defined as the mass ratio of water to the sum of soil and cement. The content of cement was 12% in the study conducted by Zhao et al. (2013). The maximum content of cement was 20% in the study conducted by Lorenzo and Bergado (2006). Comprehensive consideration of cost and actual construction needs, 50% was chosen as the maximum cement content in this study. Four mix ratios were determined as Table 1.

Fig. 1. Sampling location of Nanyang expansive soil

(a) (b) (c)

(d) (e) (f)

Fig. 2. Sample preparation processes (a) expansive soil without crushing (b) crushing and sieving (c) mixing the cement, water and soil powder (d) cemented soil samples for direct shear test (e) cemented soil samples for unconfined compressive test (f) saturating and curing

In order to simulate the real construction process of deep soil mixing method (Bhadriraju et al. 2007; Chen et al. 2013), the cement slurry was injected into the dry soil powder. As reported by Xiao et al. (2018), the dry soil powder was mixed with cement slurry in a mechanical mixer for around 10 min. Then the mixture was poured into the mold and vibrated by the vibrating machine to expel visible bubbles. The cutting ring sample was injected into two layers, and the cylindrical sample was injected into three layers. Three different curing times of 7 d, 14 d, and 28 d were conducted, and all samples were cured under constant temperature and saturation conditions. Sample preparation processes of cemented expansive soil are shown in Fig. 2.

3 Results and Discussions

Nguyen et al. (2017) indicated that the effect of cementation is diminished as the effective confining pressure is increased, due to degradation of cement-soil particle bonding. In order to consider the effects of cementation degradation during loading due to confining pressure and shearing, the Cohesion Cam Clay model (Xiao et al. 2017) should be considered as the constitutive model to simulate the behavior of cemented soils. In this study, the discussion of the constitutive model was not involved; the Mohr-Coulomb constitutive model was chosen.

Direct Shear Test

The direct shear test results are summarized in Table 1. Case 1 is the samples of expansive soil, and Case 2 to Case 13 are the samples of cemented expansive soil with different mix ratios and curing times. It should be noted that the range of dilation angles is also of significance for engineering practice. This study mainly focuses on the internal friction angle of cemented soil. The dilation angles would be analyzed in the further study. Figures 3 and 4 describe the effect of curing time and mix ratio on cohesion and internal friction angle.

Table 1. Experimental results of direct shear tests

Cases	Mix ratio (s:c:w)	Curing time (days)	Cohesion c (kPa)	Internal friction angle φ (°)
Case 1	/	/	20.10	16.64
Case 2	10:1:5.5	7	182.42	37.05
Case 3	10:1:5.5	14	285.62	36.87
Case 4	10:1:5.5	28	342.85	37.57
Case 5	5:1:3	7	206.65	49.10
Case 6	5:1:3	14	312.44	50.64
Case 7	5:1:3	28	407.56	52.14
Case 8	10:3:6.5	7	265.54	56.72
Case 9	10:3:6.5	14	387.23	55.74

(continued)

Table 1. (*continued*)

Cases	Mix ratio (s:c:w)	Curing time (days)	Cohesion c (kPa)	Internal friction angle φ (°)
Case 10	10:3:6.5	28	446.56	57.06
Case 11	10:5:7.5	7	387.17	62.32
Case 12	10:5:7.5	14	512.25	62.93
Case 13	10:5:7.5	28	587.99	63.24

Note: s = soil; c = cement; w = water

The direct shear test results are shown in Figs. 3 and 4. Figure 3 indicates that compared with the samples of expansive soils, the cohesion increases significantly after adding cement. The cemented soil of the four mix ratios has the same trend. With the increase of the curing time, the cohesion increases gradually. The cohesion increases rapidly at the beginning of curing age, and then slows down. As the results of Cases 8 to 10, the cohesion increases by 245.44 kPa after curing for 7 days, the cohesion of 14 days is 121.69 kPa larger than that of 7 days, and the cohesion of 28 days is 59.33 kPa larger than that of 14 days. Furthermore, the cohesion of cemented soil increases with the increase of cement content. After curing for 14 days, the cohesions of cemented soil are 285.62 kPa, 312.44 kPa, 387.23 kPa, and 512.25 kPa.

Figure 4 indicates that the internal friction angle of cemented soil increases significantly compared with expansive soil. However, with the increase of the curing age, the internal friction angle of cemented soil has no obvious variation; It indicates that the internal friction angle at 28 days of curing is the similar to that at seven days of curing. For the cemented soil with the same curing time, the internal friction angle increases with cement content. The internal friction angles of the cemented soil cured for 28 days are 37.57°, 52.14°, 57.06°, and 63.24° with the increase of cement content.

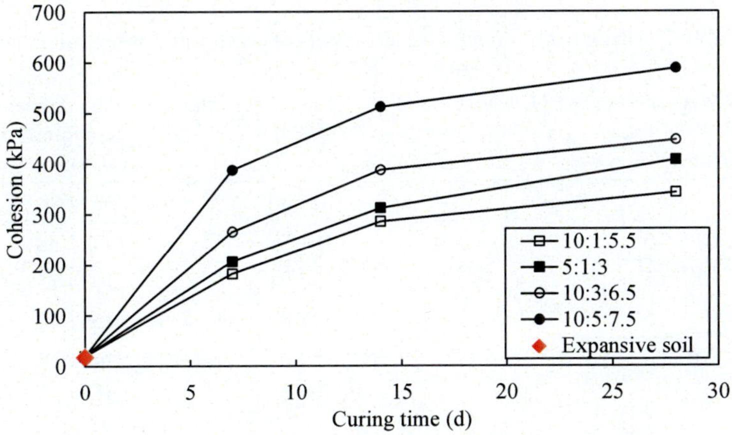

Fig. 3. Relationship among cohesion, mix ratio, and curing time

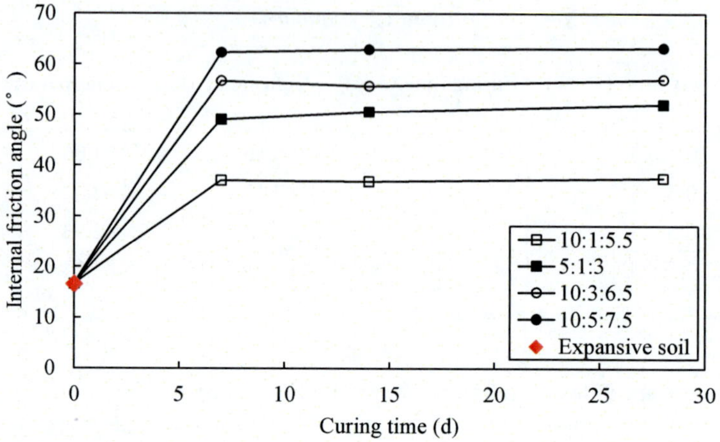

Fig. 4. Relationship among internal friction angle, mix ratio, and curing time

Unconfined Compressive Test

The unconfined compressive test results are summarized in Table 2. Figures 5 and 6 describe the effect of curing time and mix ratio on elastic modulus and unconfined compressive strength. The quantitative description of the brittleness index is also of significance for engineering practice. This study without quantitatively considering the brittleness index. This issue would be analyzed in the further study.

The elastic modulus E_{50} is taken as the slope of the line connecting the point and the origin of the axial strain corresponding to 50% of the unconfined compressive strength

$$E_{50} = \frac{q_{50}}{\varepsilon} \tag{1}$$

where E_{50}, q_{50}, and ε are elastic modulus, unconfined compressive strength and axial strain, respectively.

Table 2. Unconfined compressive test results of cases with various mix ratios

Cases	Mix ratio (s:c:w)	Curing time (days)	Elastic modulus E_{50} (MPa)	Unconfined compressive strength q_u (kPa)
Case 1	/	/	10.00	52.36
Case 2	10:1:5.5	7	161.79	947.58
Case 3	10:1:5.5	14	186.94	1374.15
Case 4	10:1:5.5	28	247.80	1592.44
Case 5	5:1:3	7	189.60	1471.99
Case 6	5:1:3	14	232.33	1952.60

(*continued*)

Table 2. (*continued*)

Cases	Mix ratio (s:c:w)	Curing time (days)	Elastic modulus E_{50} (MPa)	Unconfined compressive strength q_u (kPa)
Case 7	5:1:3	28	264.67	2264.98
Case 8	10:3:6.5	7	200.00	1902.70
Case 9	10:3:6.5	14	256.23	2611.60
Case 10	10:3:6.5	28	294.54	3224.23
Case 11	10:5:7.5	7	212.69	2617.34
Case 12	10:5:7.5	14	294.54	3750.73
Case 13	10:5:7.5	28	311.10	4442.44

Note: s = soil; c = cement; w = water

The unconfined compressive test results are shown in Figs. 5 and 6. Figure 5 indicates that compared with expansive soil, the unconfined compressive strength increases significantly after adding cement. With the increase of the curing time, the unconfined compressive strength increases gradually. The unconfined compressive strength increases rapidly at the beginning of curing age, and then slows down. This trend is the same as Tang et al. (2013). As the results of Cases 8 to 10, the unconfined compressive strength increases by 1850.34 kPa after curing for 7 days, the unconfined compressive strength of 14 days is 708.90 kPa larger than that of 7 days, and the unconfined compressive strength of 28 days is 612.63 kPa larger than that of 14 days. With the increase of cement content, the unconfined compressive strength of cemented soil increases. After curing for 14 days, with the increase of cement content, the unconfined compressive strengths of cemented soil are 1374.15 kPa, 1952.60 kPa, 2611.60 kPa, and 3750.73 kPa.

Figure 6 indicates that the elastic modulus of cemented soil gradually becomes more extensive as the curing time increases, which is significantly improved compared to the 10 MPa of expansive soil. For the cemented soil with the same curing time, the elastic modulus increases with the cement content. However, the elastic modulus of cemented soil with a large amount of cement does not increase significantly. Taking the cemented soil after curing for 28 days as an example, as the cement content increases, the elastic moduli of cemented soil are 247.80 MPa, 264.67 MPa 294.54 MPa, and 311.10 MPa.

Failure Mode

According to the test results, the failure mode of cemented soil is different from that of expansive soil (Fig. 7). The failure mode of expansive soil is a plastic failure, while cemented soil has a brittle failure. It can be found from Fig. 7(d) that the brittle failure becomes severe as the increase of cement content. The phenomenon of fracture existing during the failure process of cemented soil, while the samples of expansive soil have no obvious fracture.

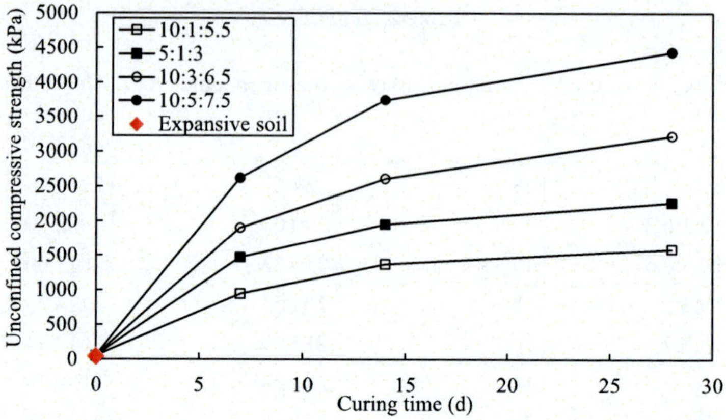

Fig. 5. Relationship among unconfined compressive strength, mix ratio, and curing time

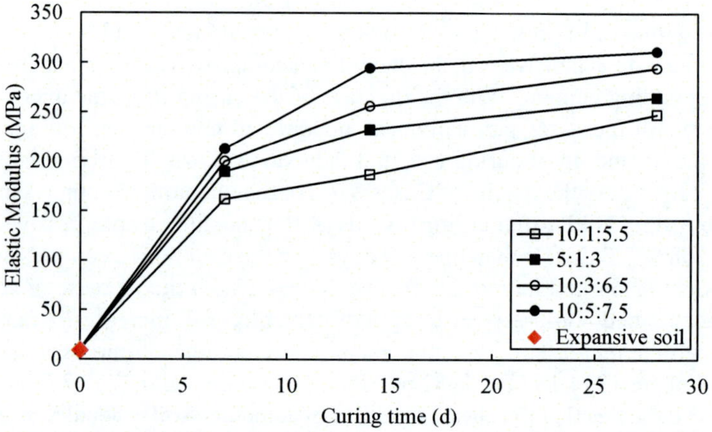

Fig. 6. Relationship among elastic modulus, mix ratio, and curing time

Discussions

Cement has a high affinity for water and short setting time. The hydration reaction of cement cements the expansive soil particles, and the mechanical properties of the expansive soil are improved. The rapid increase at the beginning of curing age is because the hydration reaction is violent in the early stage, and the pozzolanic reaction tends to be gentle in the later stage. Pozzolanic reaction can form a new insoluble compound, which hardens when cured and thereby stabilizes the soil (Lorenzo and Bergado 2006). Pozzolanic reactions become severe as the increase of cement content (Nelson and Miller 1992; Punthutaecha 2002), so the brittle failure becomes severe.

During the sample preparation process, the water content was fixed at 50%. However, with the increase of the cement content, the liquid limit of cemented soil decrease. Cemented soil with high cement content is easier to reach the bleeding limit, so elastic

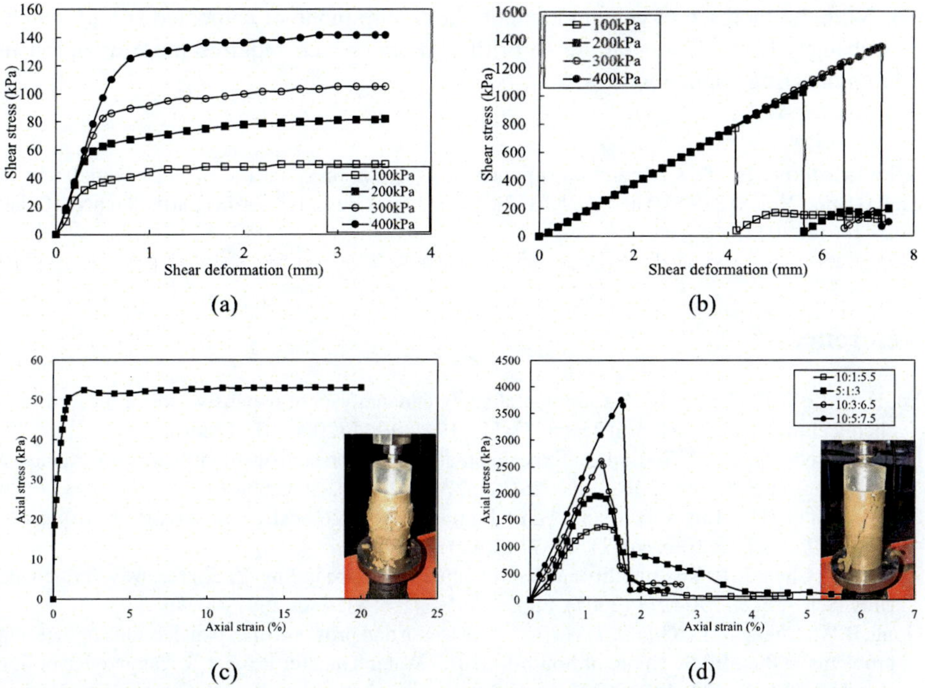

Fig. 7. Stress-strain relationship of (a) expansive soil in direct shear test (b) cemented soil in direct shear test (Case 13); (c) expansive soil in unconfined compressive test (d) cemented soil in unconfined compressive test (Case 3, Case 6, Case 9, and Case 12)

modulus of the cemented soil with a large amount of cement does not increase significantly in this study. Although the unconfined compressive strength and shear strength of cemented soil increase with the cement content, the failure mode should be considered to prevent severe consequences in engineering applications.

4 Conclusion

Relationship among mechanical properties and cement content and curing time were analyzed qualitatively and quantitatively through laboratory investigation. The following conclusions can be drawn from this investigation.

(1) Compared with expansive soil, the mechanic properties of the cemented soil have been improved significantly after stabilized by cement.
(2) With the increase of the curing time, the cohesion of the cemented soil increases gradually, while the internal friction angle of cemented soil has no obvious variation. The cohesion increases rapidly at the beginning of curing age, and then slows down.
(3) The cohesion and elastic modulus of cemented soil gradually increase with the curing time. It both increases significantly at the beginning of curing, and slower in the later stage. The cohesion and elastic modulus increase with cement content.

(4) With the increase of cement content, the failure mode of cemented soil gradually changes from plastic failure to brittle failure, which requires more attention in engineering applications.

Acknowledgments. This research is supported by the National Natural Science Foundation of China (Grant No.: 51879203) and the NRF-NSFC 3rd Joint Research Grant (Earth Science) (Grant No.: 41861144022).

References

Yu, H.Z., Li, X.Q., Yao, J.W.: Experimental study and analysis of expansive soil improved with chemical medicine. Rock Soil Mech. **27**(11), 1941–1944 (2006). (in Chinese)

Huang, B., Nie, Q., Xu, Y.Y., Luo, J.: Experimental research on cement-modification of expansive Soil. J. Yangtze River Sci. Res. Inst. **26**(11), 27–30 (2009). (in Chinese)

Shi, B., Jiang, H.T., Liu, Z.B., Fang, H.Y.: Engineering geological characteristics of expansive soils in China. Eng. Geol. **67**(1), 63–71 (2002)

Zuo, C.-Q., Chen, J.-P.: The stability analysis of expansive slope in Jing-Yi Expressway. J. Zhejiang Univ. Sci. A **8**(2), 270–274 (2007). https://doi.org/10.1631/jzus.2007.A0270

Gong, B.W., Cheng, Z.L., Guo, X.L., Li, Q.Y.: Research and progress of expansive soil engineering problems in the Middle Route of South-to-North Water Transfer Project. J. Yangtze River Sci. Res. Inst. **28**(10), 134–140 (2011). (in Chinese)

Tang, Y.W., Tang, L., Zhang, G.D.: Test of the unconfined compressive strength of improved cement expansive soil. J. Huaiyin Inst. Technol. **22**(3), 26–30 (2013). (in Chinese)

Madhyannapu, R.S., Puppala, A.J., Bhadriraju, V.: Deep soil mixing (DSM) treatment of expansive soils. Adv. Ground Improve. **3**, 130–139 (2019)

Pora, S., Nishimurab, S., Likitlersuangc, S.: Deformation characteristics and stress responses of cement-treated expansive clay under confined one-dimensional swelling. Appl. Clay Sci. **146**, 316–324 (2017)

Lee, F.H., Lee, Y., Chew, S.H., Yong, K.Y.: Strength and modulus of marine clay-cement 170 mixes. J. Geotech. Geoenviron. Eng. **131**(2), 178–186 (2005)

Xiao, H.W., Yao, K., Liu, Y., Goh, S.A., Lee, F.H.: Bender element measurement of small strain shear modulus of cement-treated marine clay – effect of test setup and methodology. Constr. Build. Mater. **173**, 433–447 (2018)

Khemissa, M., Mohanmedi, A.: Cement and lime mixture stabilization of an expansive overconsolidated clay. Appl. Clay Sci. **95**, 104–110 (2014)

Nguyen, L., Fatahi, B., Khabbaz, H.: Development of a constitutive model to predict the behavior of cement-treated clay during cementation degradation: C3 model. Int. J. Geomech. **17**(7), 1–32 (2017)

Xiao, H.W., Lee, F.H., Liu, Y.: Bounding surface cam-clay model with cohesion for cement-admixed clay. Int. J. Geomech. **17**(1), 1–22 (2017)

Bhadriraju, V., Puppala, A.J., Madhyannapu, R.S., Williammee, R.: Laboratory procedure to obtain well-mixed soil binder samples of medium stiff to stiff expansive clayey soil for deep soil mixing simulation. Geotech. Test. J. **31**, 225–238 (2007)

Chen, J.J., Zhang, L.Y., Zhang, J.F., Zhu, Y.F., Wang, J.H.: Field tests, modification, and application of deep soil mixing method in soft clay. J. Geotech. Geoenviron. Eng. **33**, 24–34 (2013)

Nelson, J.D., Miller, J.D.: Expansive soils: Problems and Practice in Foundation and Pavement Engineering. Wiley, New York (1992)

Punthutaecha, K.: Volume change behavior of expansive soils modified with recycled materials. Doctoral thesis, University of Texas (2002)
Lorenzo, G.A., Bergado, D.T.: Fundamental characteristics of cement-admixed clay in deep mixing. J. Mater. Civ. Eng. **16**, 161–172 (2006)
Zhao, H.H., Zhao, C.J., Zheng, X., Liu, C.: Physicochemical characterization of cement stabilized highly expansive soil. In: Geo-Congress, pp. 2091–2099 (2013)
Shen, S.L., Hong, Z.S., Xu, Y.S.: Reducing differential settlements of approach embankments. Geotech. Eng. **4**, 215–226 (2007)
Ye, G.B., Chang, H.T., Xu, C., Liao, X.Y.: Field tests, modification, and application of deep soil mixing method in soft clay. In: Contemporary Topics in Ground Modification, Problem Soils, and Geo-Support, pp. 209–216 (2009)
Madhyannapu, R.S.: Deep mixing technology for mitigation of swell-shrink behavior of expansive soils moderate to deep active depths. Doctoral thesis, University of Texas (2006)
Madhyannapu, R.S., Puppala, A.J., Williammee, R.: Design and construction guidelines for deep soil mixing to stabilize expansive soils. J. Geotech. Geoenviron. Eng. **140**(9), 1–15 (2014)

states of MICP treatment (Lin et al. 2016; Xiao et al. 2019d; Xiao et al. 2020). Although these technologies produce some insights into the microstructure of biotreated samples, they cannot capture the dynamic precipitation of calcium carbonate in pore space. This study provides a novel technique by taking advantage of microfluidics to visualize the process of crystal growth and the formation of calcium carbonate bonding between sand particles under flow condition. Preliminary results are presented and discussed, which could help understand the growth behavior of calcium carbonate at the pore-scale.

2 Materials and Methods

2.1 Conceptual Microchip

Continuous pore space of subsurface materials mainly consists of large voids and pore throats. An idealized microfluidic chip is designed to mimic this geometric feature of porous media. The microfluidic chip consists of a main channel and sand particles inside. The pore throats are obtained by placing the sand particles into a row, while the large voids are the remaining spaces apart from sand particles. The width of the main channel is 950 μm, the depth is 120 μm. The width of pore throats ranges from 30–130 μm. The solutions are injected through two inlet ports and discharged from an outlet port as shown in Fig. 1(a). The diameters of inlet and outlet ports are slightly larger than that of the main channel to avoid possible reflux. See a small portion of the outlet port at the right side of Fig. 1(b). The microchip is fabricated by polydimethylsiloxane using standard photolithography method. Interested readers could refer to Xia and Whitesides (1998) and Wang et al. (2019) for more details.

2.2 Reaction Solutions

Sporosarcina pasteurii (ATCC 11859) is used in this study. The culture medium and cultivation methods of this bacterium could be found in previous researches. Cultures are centrifuged at 7000g for 15 min at 4 °C and resuspended in 0.85% NaCl solution. Cementation solution (CS) is composed of 0.5 mol/L urea and 0.5 mol/L $CaCl_2$. The composition of culture medium and cementation solutions is detailed in Table 1.

2.3 $CaCO_3$ Precipitation and Visualization

The bacterial and cementation solutions are injected into the main channel through two separate tubes as shown in Fig. 1(a). The tubes are connected with syringe pumps (Harvard Elite 11, USA) which offer smooth and constant flow. The solutions are mixed in the microchannel and precipitation occurs. The formation of $CaCO_3$ could take place in large voids or pore throats among sand particles. This process is continuously monitored by an inverted microscope with a 10× objective and captured by a digital camera (Olympus DP74, Japan). The imaging area is 6 mm away from the intersection of the two solutions.

Table 1. Composition of solutions

Injection solutions	Chemical	Chemical concentration
Bacterial solution	Yeast extract	20 g/L
	NH_4Cl	10 g/L
	$MnSO_4{\cdot}2H_2O$	12 mg/L
	$NiCl_2{\cdot}6H_2O$	24 mg/L
Cementation solution	$CaCl_2$	55.49 g/L
	$CO(NH_2)_2$	30.03 g/L

2.4 Image Processing

The raw images are used to distinct the sand particles, the microchannel, and crystals. The identified crystals, sand particles are then filled with different colors to present the growth states at different reaction time.

3 Test Results and Discussion

Crystals formed in pore throats and large voids have different shapes as shown in Fig. 1(b). Irregular aggregates are observed in the vicinity of sand particles, while regular and polyhedral crystals are noted in large voids. This phenomenon indicates that the precipitation behaviors might be different in different pore structures. The precipitates will be quantitatively analyzed in the following sections.

3.1 Growth of Single Crystal

Crystal formed in the middle of the microchannel, as shown in the dashed box in Fig. 1(b), is picked for single crystal analysis. The raw images of this isolated crystal are shown in Figs. 2(a)–(d), the corresponding processed images are presented in Figs. 2(e)–(h). The white small arrows represent the direction of the flow, while the big vertical arrow indicates the elapsing of reaction time. It is noted that this crystal is not as symmetrical or rhombohedral as those typically found in chemically-induced $CaCO_3$. Meanwhile, the crystal presents a preferential growth at the upstream face as shown in Figs. 2(h), indicating that the flow direction affects the growth of $CaCO_3$ in the MICP process.

The growth behavior of the single crystal is shown in Fig. 2(i) and (j). The size of the crystal displays a progressive increase with the reaction time (see Fig. 2(i)). A similar shape could be observed for this crystal after a certain size (around 0.038 μm^2) is reached. The growth rate increases at the beginning, then becomes almost constant, and decreases at around the 990th min as shown in Fig. 2(j). The most rapid growth occurs at the 607th min with an approximate rate of 55 μm^2/min.

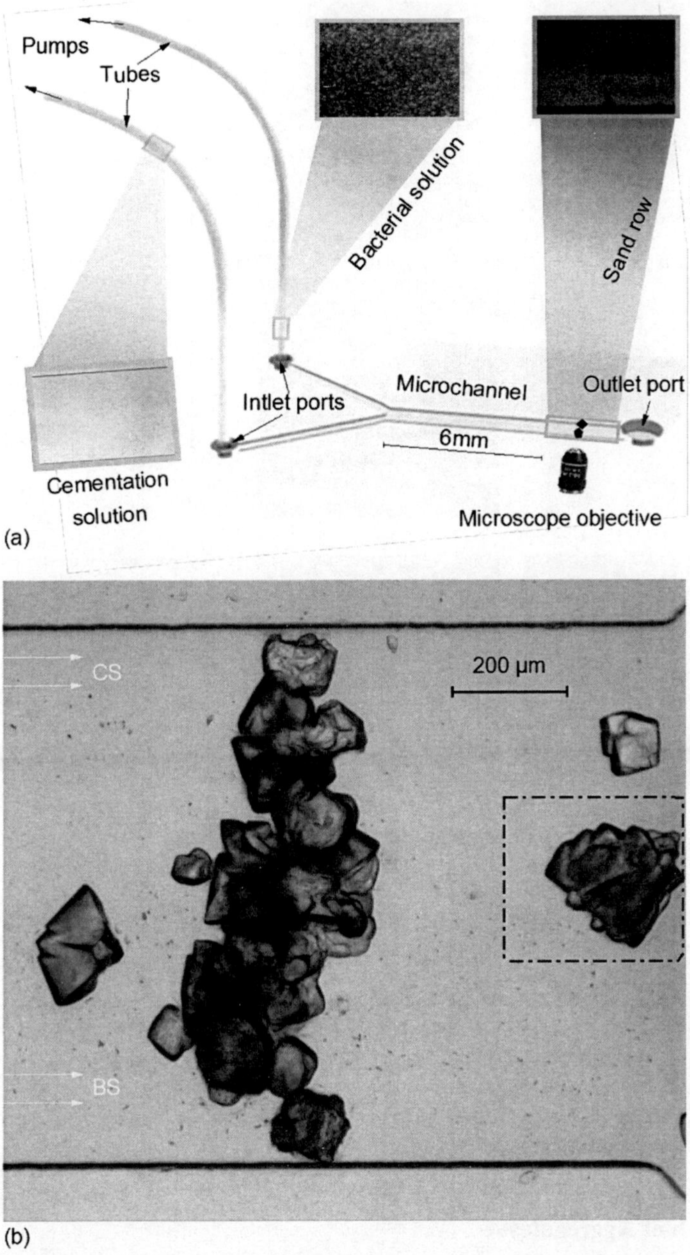

Fig. 1. (a) Schematic of MICP microfluidic tests, (b) raw image at the end of the test.

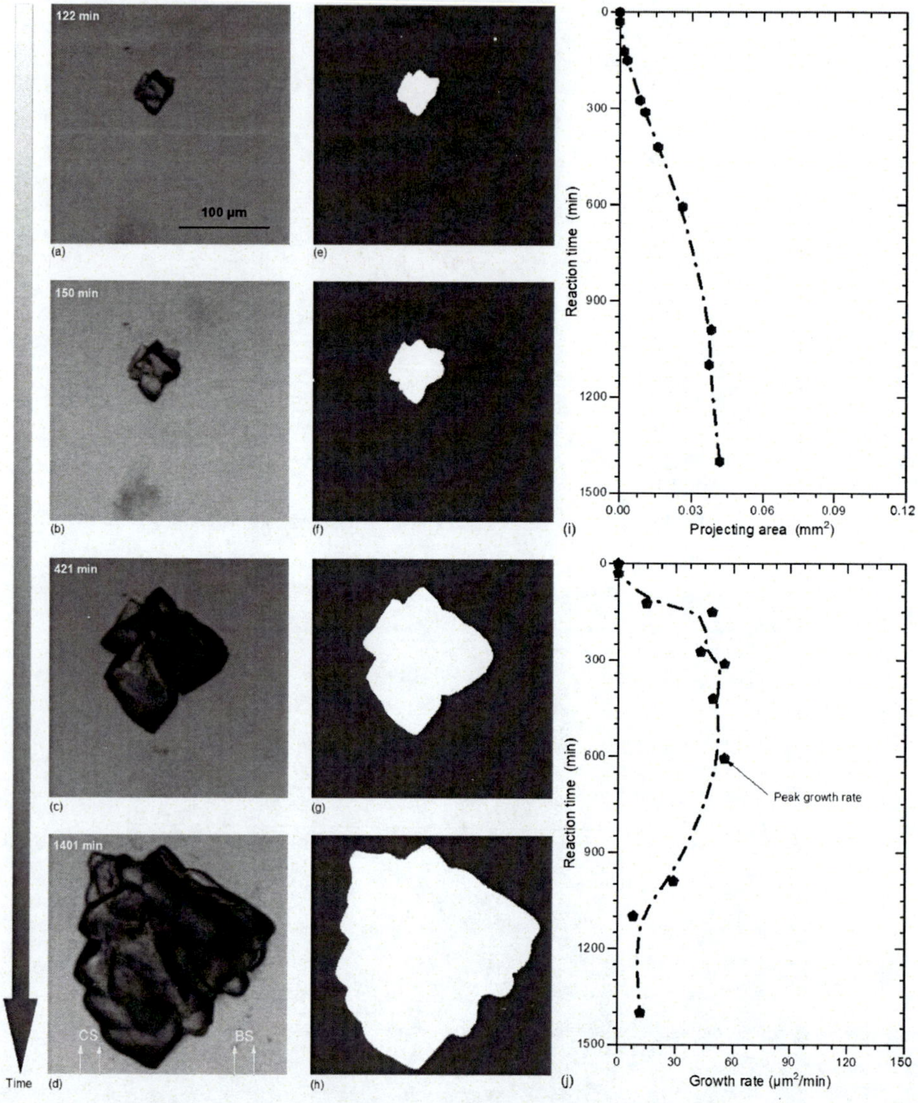

Fig. 2. Single crystal in terms of: evolution in (a)–(d) raw images and (e)–(h) processed images, (i) accumulative projecting area and (j) growth rate.

3.2 Growth of Aggregates

As shown in Fig. 3, the crystals in the vicinity of sand grains appear first in the BS side and then in the CS side of the microchannel. They gradually enwrap the sand particles as they grow bigger. The accumulative behaviors of crystals are quantitatively analyzed and the results are shown in Fig. 3(i) and (j). Similar to the single crystal, the crystalline aggregates accumulate as the reaction time increases. The projecting area of aggregates even exceed that of sand grains at the 1167th min. The growth rate reaches to its peak at

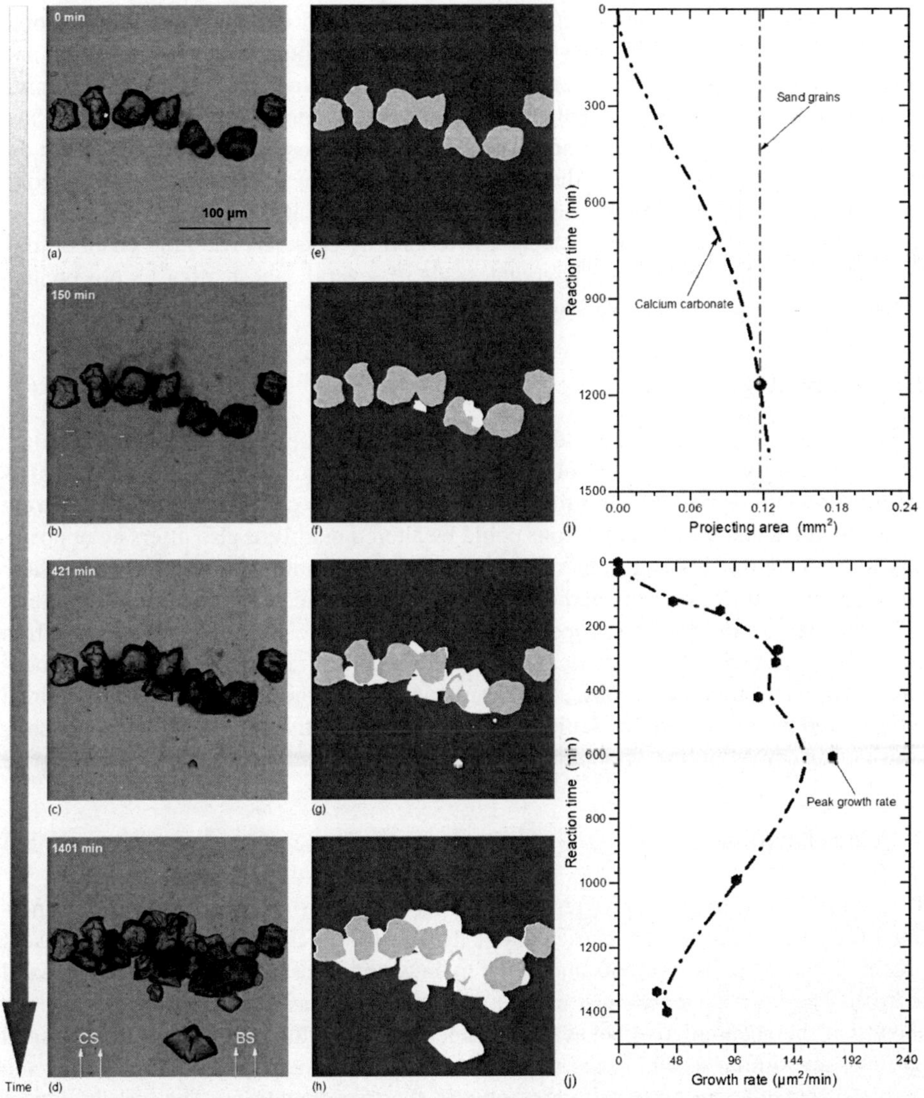

Fig. 3. Crystalline aggregates in terms of: evolution in (a)–(d) raw images and (e)–(h) processed images, (i) accumulative projecting area and (j) growth rate.

the 607th min (see Fig. 3(j)), which is the same as that for the single crystal. However, the difference between the upstream and the downstream are not obvious for crystalline aggregates as compared with that for the single crystal. This phenomenon implies that the growth behaviors of crystals in large voids are different from those in pore throats of sand particles. In contrast to the laminar flow in the main channel, the fluid flow across sand particles are more tortuous, which leads to a relatively homogeneous growth of aggregates.

The evolution of precipitation around particles could be characterized into three stages according to Fig. 3, namely surface precipitation, inter-particle cementation and channel clogging. When the content of $CaCO_3$ was relatively low, i.e. at the beginning of reaction as present in Figs. 3(b) and 3(f), the crystals formed on the particle surface without effective interparticle bonds. The formation of those crystals may increase the roughness of particle surface. As the content of $CaCO_3$ increased, crystals formed among sand particles, as shown in Figs. 3(c) and 3(g), leading to inter-particle cementation. At the third stage, the sand particles were enwrapped by crystals. The fluid channel may be clogged by the sand-crystal aggregates where several sand particles were bonded together as shown in Figs. 3(d) and 3(h).

3.3 Potential Implications

This work presents a tool that could visualize the precipitation process of biomineralization. The primary results herein validate the feasibility to quantify the growth of crystals among the sand matrix using this tool. Since the microchip could be tailored to different pore structures and the test conditions could be altered readily, it also offers an opportunity to evaluate the effects of other factors, such as the solution concentration, the flow rate, injection strategies, etc., and to optimize the practical MICP scheme in the field. Furthermore, the mechanical properties of the interparticle cementation formed herein could be examined with a specifically-designed loading framework, while the cementation degradation could be observed simultaneously at the microscale. The obtained information would provide valuable reference for constitutive models of cemented soils considering cementation degradation under compression and shear (Nguyen et al. 2017).

4 Conclusions

This study investigates the progressive growths of $CaCO_3$ under different flow conditions (i.e., flow across sand particles and flow in the microchannel) which represent precipitation in the pore throats and in the main channel (or large voids in the real sand matrix). The raw images are processed to evaluate the precipitation behaviors. Results show that the accumulative behaviors of precipitation in the pore throats among sand particles are similar to that in the main microchannel in the present test. However, the shapes of crystals are quite different between the two conditions. The distribution of crystals is more uniform for crystals precipitated in the vicinity of sand particles, while a preferential growth at the upstream face is noted for the single crystal in the main microchannel.

Acknowledgments. The authors would like to acknowledge the financial support from the National Natural Science Foundation of China (Grant No. 51678094 and Grant No. 51922024), the Natural Science Foundation of Chongqing, China (Grant No. cstc2019jcyjjqX0014) and the Graduate Research and Innovation Foundation of Chongqing, China (Grant No. CYB18035).

References

Al Qabany, A., Soga, K.: Effect of chemical treatment used in micp on engineering properties of cemented soils. Géotechnique **63**(4), 331–339 (2013)

Cheng, L., Shahin, M.A., Mujah, D.: Influence of key environmental conditions on microbially induced cementation for soil stabilization. J. Geotech. Geoenvironmental Eng. **143**(1), 11 (2017)

Chu, J., Ivanov, V., Stabnikov, V., Li, B.: Microbial method for construction of an aquaculture pond in sand. Geotechnique **63**(10), 871–875 (2013)

Clarà Saracho, A., Haigh, S.K., Ehsan Jorat, M.: Flume study on the effects of microbial induced calcium carbonate precipitation (micp) on the erosional behaviour of fine sand. In: Géotechnique, pp. 1–15 (2020)

DeJong, J.T., Mortensen, B.M., Martinez, B.C., Nelson, D.C.: Bio-mediated soil improvement. Ecol. Eng. **36**(2), 197–210 (2010)

Lin, H., Suleiman, M.T., Brown, D.G., Kavazanjian, E.: Mechanical behavior of sands treated by microbially induced carbonate precipitation. J. Geotech. Geoenvironmental Eng. **142**(2), 04015066 (2016)

Liu, L., Liu, H., Stuedlein, A.W., Evans, T.M., Xiao, Y.: Strength, stiffness, and microstructure characteristics of biocemented calcareous sand. Can. Geotech. J. **56**(10), 1502–1513 (2019)

Martinez, B.C., et al.: Experimental optimization of microbial-induced carbonate precipitation for soil improvement. J. Geotech. Geoenvironmental Eng. **139**(4), 587–598 (2013)

Mitchell, J.K., Santamarina, J.C.: Biological considerations in geotechnical engineering. J. Geotech. Geoenvironmental Eng. **131**(10), 1222–1233 (2005)

Nguyen, L., Fatahi, B., Khabbaz, H.: Development of a constitutive model to predict the behavior of cement-treated clay during cementation degradation: C3 model. Int. J. Geomech. **17**(7), 04017010 (2017)

Soon, N.W., Lee, L.M., Khun, T.C., Ling, H.S.: Factors affecting improvement in engineering properties of residual soil through microbial-induced calcite precipitation. J. Geotech. Geoenvironmental Eng. **140**(5), 04014006 (2014)

Wang, Y.Z., Soga, K., Dejong, J.T., Kabla, A.: A microfluidic chip and its use in characterising the particle-scale behaviour of microbial-induced calcium carbonate precipitation (micp). Geotechnique **69**(12), 1086–1094 (2019)

Xia, Y.N., Whitesides, G.M.: Soft lithography. Ann. Rev. Mater. Sci. **28**, 153–184 (1998)

Xiao, P., Liu, H.L., Stuedlein, A.W., Evans, T.M., Xiao, Y.: Effect of relative density and biocementation on cyclic response of calcareous sand. Can. Geotech. J. **56**(12), 1849–1862 (2019a)

Xiao, P., Liu, H.L., Xiao, Y., Stuedlein, A.W., Evans, T.M.: Liquefaction resistance of bio-cemented calcareous sand. Soil Dyn. Earthq. Eng. **107**, 9–19 (2018)

Xiao, Y., et al.: Restraint of particle breakage by biotreatment method. J. Geotech. Geoenvironmental Eng. **146**(11), 04020123 (2020). https://doi.org/10.1061/(ASCE)GT.1943-5606.0002384

Xiao, Y., He, X., Evans, T.M., Stuedlein, A.W., Liu, H.: Unconfined compressive and splitting tensile strength of basalt fiber-reinforced biocemented sand. J. Geotech. Geoenvironmental Eng. **145**(9), 04019048 (2019b)

Xiao, Y., et al.: Effect of particle shape on strength and stiffness of biocemented glass beads. J. Geotech. Geoenviron. Eng. **145**(11), 06019016 (2019c)

Xiao, Y., Wang, Y., Desai, C.S., Jiang, X., Liu, H.L.: Strength and deformation responses of biocemented sands using a temperature-controlled method. Int. J. Geomechanics **19**(11), 10 (2019d)

caused by consolidation is large, which makes the small strain assumption invalid. To solve this problem, large strain consolidation theory was built to consider the variation of soil thickness. Gibson et al. (1981) proposed nonlinear large strain theory for highly compressible soils at first. After that, studies of soil consolidation based on large strain assumptions have been developed. Hu et al. (2014) developed a large strain elastic viscoplastic consolidation model for soil improved by PVDs. Geng et al. (2017) presented a 2D nonlinear large strain consolidation model under vacuum pressure with both vertical and radial flow considered. In recent years, the radial large strain consolidation has been expanded to take the non-Darcy flow law (Indraratna et al. 2017; Pu et al. 2020) and well resistance (Nguyen et al. 2020) into consideration.

Recently, several studies have reported that the consolidation of the dredged slurries was quite nonuniform (Fang et al. 2019). Soil columns formed surrounding the vertical drains with much higher strength and lower water content than the outside. In contrast, the soils at the periphery of the PVDs seems unimproved, where the water content in this part was still high, and bearing capacity was still low (Liu et al. 2017; Lei et al. 2017). This phenomenon was explained by the clogging of the vertical drains and the soil adjacent to the vertical drains (Bao et al. 2014a, b). As fine particles were found rich in dredged slurries (Deng et al. 2019), they were driven to the drain under the hydraulic gradient in the process of improvement, which blocked the drainage channel of the drains and void in the soil near the drain. Then, the water in the slurries cannot be drained out (Tang et al. 2010). Experiments of the vacuum distribution along the drains showed, different from the ideal uniform distribution, the vacuum pressure in practice varies with distance to the vacuum connector at the drain (Bao et al. 2014a, b), and increases from zero to the target value gradually. However, the aforementioned large strain consolidation theories cannot reflect these factors occurring in the consolidation of dredged slurries.

In this study, a consolidation model is built to characterize the consolidation of dredged slurries by considering vacuum pressure depending on the time and distance to the vacuum connector at the boundaries. FEM computations were made to investigate the influence of time variation and distance variation on the consolidation behavior. Comparisons are made between the small strain solutions and the large strain solution to show the difference of the influence. Finally, a case study is conducted to investigate the capability of the model.

2 Model Description

Vacuum pressure has been observed to decrease with depth in many laboratory and field tests (Cai et al. 2017; Bao et al. 2014a). In the improvement of dredged slurries, the vacuum loss is found not only existing in the vertical drain but also the radial drainage path from the horizontal drain to the membrane. The vacuum pressure under the membrane is smaller than that in the horizontal drain. Moreover, the vacuum pressure increases gradually in the drain of both directions and under the membrane. Therefore, to consider the actual distribution of the vacuum pressure in the improvement of slurries, boundary conditions at the vertical and horizontal drains were suggested in the form of Eqs. (1) and (2) (Bao et al. 2014b)

$$p|_{z=0} = p_0 f(r) f(t) \tag{1}$$

$$p|_{r=r_w} = p_0 g(z) g(t) \tag{2}$$

where p_0 is the value of vacuum pressure sent by the pump, r is the distance from the center of PVD, z is the distance from the surface of the soil, t is the elapsed time, $f(r)$ is the function to describe the radial distribution, $f(t)$ is the function to describe the variation with time in the radial direction, $g(z)$ is the function to describe the vertical distribution along the drain, and $g(t)$ is the function to describe the variation with time in the vertical direction.

By fitting vacuum measured in laboratories and fields, the vacuum pressure's reduction along depth and radius can be depicted in linear form, and the variation with time can be simulated in exponential form (Liu et al. 2019). Therefore, Eqs. (1) and (2) can be rewritten as

$$\begin{cases} p|_{z=0} = p_0\left(1 - (1 - k_{r1})\frac{(r - r_w)}{(r_e - r_w)}\right)\left(1 - e^{-k_{r2}t}\right) \\ p|_{r=r_w} = p_0\left(1 - (1 - k_{z1})\frac{z}{H}\right)\left(1 - e^{-k_{z2}t}\right) \end{cases} \tag{3}$$

where k_{r1} is the coefficient for radial distribution, k_{r2} is the coefficient for time variation in the radial direction, k_{z1} is the coefficient for vertical distribution, k_{z2} is the coefficient for time variation in the vertical direction, r_w is the equivalent radius of the vertical drain; r_e is the equivalent radius of drain-influenced zone and H is the length of vertical drain.

Because of clogging in the soil, a special zone adjacent to the drain (also called soil column) is investigated in the soil, where permeability and compressibility are different from outside. It is called clogging zone below. This area forms because of the immigration of fine particles and nonuniform consolidation. The scheme of the axisymmetric unit cell shown in Lagrangian coordinates is used here to simulate the effective range of soil surrounding a single drain as shown in Fig. 1, where a is the Lagrangian coordinate, r_c is the radius of clogging zone, k_c and k_h are the radial permeable coefficients in the clogging zone and normal zone, respectively, k_r is the permeable coefficient at any radius, k_v is the vertical permeable coefficient.

To solve the problem easily, the reduced coordinate (z) is used in the governing equation. Combined the relationship between the reduced coordinate and the Lagrangian coordinate with the governing equation in Geng et al. (2017), the governing equation about e in the reduced coordinate can be expressed as below

$$\begin{aligned} &\frac{1}{r}\frac{\partial}{\partial r}\left[r\frac{k_c(e)}{\gamma_w}\frac{d\sigma'}{de}\frac{\partial e}{\partial r}\right] + \frac{1}{(1+e)}\frac{\partial}{\partial z}\left[\frac{k_v(e)}{\gamma_w}\frac{d\sigma'}{de}\frac{\partial e}{\partial z}\frac{1}{1+e} + \frac{k_v(e)(G_s - 1)}{1+e}\right] = -\frac{1}{1+e}\frac{\partial e}{\partial t} \quad (r \le r_c) \\ &\frac{1}{r}\frac{\partial}{\partial r}\left[r\frac{k_r(e)}{\gamma_w}\frac{d\sigma'}{de}\frac{\partial e}{\partial r}\right] + \frac{1}{(1+e)}\frac{\partial}{\partial z}\left[\frac{k_v(e)}{\gamma_w}\frac{d\sigma'}{de}\frac{\partial e}{\partial z}\frac{1}{1+e} + \frac{k_v(e)(G_s - 1)}{1+e}\right] = -\frac{1}{1+e}\frac{\partial e}{\partial t} \quad (r > r_c) \end{aligned} \tag{4}$$

where σ' is the effective stress, $k_r(e)$, $k_c(e)$ and $k_v(e)$ are horizontal permeable coefficients varying with void ratio in the normal zone, clogging zone and in the vertical direction, respectively, G_s is the specified gravity. The relationship between the permeable coefficients and the void ratio, and the relationship between the effective stress and

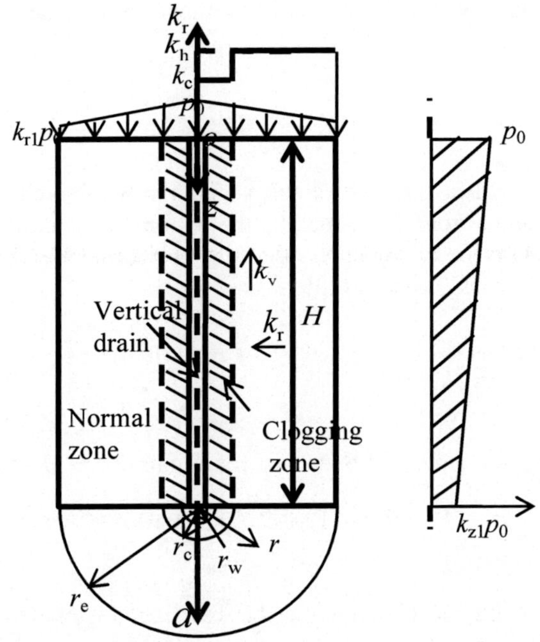

Fig. 1. Analysis scheme of unit cell with vertical drain

void ratio have been proposed in various forms. Here, the widely used e-log$k$ and e-logσ' relationships are chosen in the following cases, which can be expressed as

$$\begin{aligned} k_r &= k_{r0} 10^{\frac{e-e_{0,0}}{C_{kh}}} \\ k_c &= k_{c0} 10^{\frac{e-e_{0,0}}{C_{kh}}} \\ k_v &= k_{v0} 10^{\frac{e-e_{0,0}}{C_{kv}}} \\ \frac{d\sigma'}{de} &= \sigma'_{0,0} \frac{\ln 10}{-C_c} 10^{\frac{e-e_{0,0}}{-C_c}} \end{aligned} \tag{5}$$

where C_c is the compressive index, C_{kh} and C_{kv} are the radial and vertical permeability index, respectively; k_{r0}, k_{c0} and k_{v0} are the radial and vertical permeability coefficients of soils at the initial time, and $e_{0,0}$ and $\sigma'_{0,0}$ is the void ratio and effective stress at $a = 0$ when $t = 0$.

Since drain boundary and the upper surface of the soil are assumed pervious, the excess pore water pressure at the boundaries is equal to the vacuum pressure expressed by Eq. (3). Correspondingly, the void ratio at these two boundaries can be expressed as below:

$$\begin{cases} e|_{z=0} = e_{0,0} - C_c \log_{10}\Big(\big(\sigma'_{0,0} - p_0(1-(1-k_{r1})(r-r_w)/(r_e-r_w))(1-\exp(-k_{r2}t))\big)/\sigma'_{0,0}\Big) \\ e|_{r=r_w} = e_{0,0} - C_c \log_{10}\Big(\big(\sigma'_{0,0} - p_0(1-(1-k_{z1})z/H)(1-\exp(-k_{z2}t)) - \gamma_w(G_s-1)z\big)/\sigma'_{0,0}\Big) \end{cases} \tag{6}$$

The outer boundary of the effective zone is impervious because of the symmetry, which can be expressed as

$$\left.\frac{\partial e}{\partial r}\right|_{r=r_{\mathrm{e}}} = 0 \tag{7}$$

If the bottom of soil is assumed pervious, vacuum pressure at that boundary will be reduced along the radius from the vacuum at the bottom of the drain. The excess pore water pressure equals to the vacuum pressure at the pervious boundary, so the boundary condition at the bottom can be expressed as

$$e|_{z=H} = e_{0,0} - C_{\mathrm{c}}\log_{10}\left(\begin{pmatrix}\sigma_{0,0} - p_0(1-(1-k_{\mathrm{z1}}))(1-\exp(-k_{\mathrm{z2}}t))(1-(1-k_{\mathrm{r1}})(r-r_{\mathrm{w}})/(r_{\mathrm{e}}-r_{\mathrm{w}})) \\ -\gamma_{\mathrm{w}}(G_{\mathrm{s}}-1)H\end{pmatrix}/\sigma_{0,0}\right) \tag{8}$$

If the bottom is impervious, the boundary conditions at $z = H$ should be

$$\left.\frac{\partial e}{\partial z}\right|_{z=H} = \frac{(G_{\mathrm{s}}-1)\gamma_w C_{\mathrm{c}}/\ln(10)10^{((e-e_0)/C_{\mathrm{c}})}}{\sigma_0'} \tag{9}$$

The boundary conditions at the interface between the clogging zone and the normal zone can be expressed as

$$k_{\mathrm{c}}\left.\frac{\partial e}{\partial r}\right|_{r_{\mathrm{c}}^-} = k_{\mathrm{h}}\left.\frac{\partial e}{\partial r}\right|_{r_{\mathrm{c}}^+},\quad e|_{r_{\mathrm{c}}^-} = e|_{r_{\mathrm{c}}^+} \tag{10}$$

Since the governing equation is highly nonlinear, finite element method is used here to solve it in the following cases.

3 Model Performance

In this part, several cases are calculated to investigate the influence of vacuum pressure variation on the consolidation. Values of basic parameters for the analysis are: $r_{\mathrm{w}} = 0.07$ m, $r_{\mathrm{c}} = 3r_{\mathrm{w}}$, $r_{\mathrm{e}} = 0.7$ m, $H = 10$ m, $C_{\mathrm{c}} = 0.4$ m, $e_{0,0} = 1.1$ m, $\sigma_{0,0}' = 20$ kPa, $k_{\mathrm{r0}} = 4.48 \times 10^{-4}$ m/day, $k_{\mathrm{c0}} = 1/10k_{\mathrm{r0}}$, $k_{\mathrm{v0}} = 2.27 \times 10^{-4}$ m/day, $p_0 = -80$ kPa. According to Geng's work (2017), the consolidation degree in the form of pore water pressure can be expressed as

$$U_{\mathrm{p}} = \frac{\int_0^H \int_{r_w}^{r_{\mathrm{e}}} 2\pi r u \mathrm{d}r \mathrm{d}z}{p_0\pi\left(r_{\mathrm{e}}^2 - r_{\mathrm{w}}^2\right)H} \tag{11}$$

Excess pore water pressure can be expressed as

$$u = \sigma_0' + q - (G_{\mathrm{s}}-1)z\gamma_{\mathrm{w}} - \sigma_0 10^{(e_0-e)/C_{\mathrm{c}}} \tag{12}$$

Effect of Time Variation

By fitting the previous laboratory and field measurements (Liu et al. 2019), the time-dependent factor k_{r2} and k_{z2} varied from 0.4 to 1.1. The loss of vacuum pressure along depth and radius can be up to 57% (Bao et al. 2014b). Here a more extensive range of k_{r2} and k_{z2} is applied to study the influence of time variation. As shown in Fig. 2, the vacuum pressure develops faster with larger k_{r2} and k_{z2}. Comparing the variations for $k_{r2} = 0.01$, 0.2 and 0.4, the speed of vacuum pressure increases nonlinearly with the increase of k_{r2}. When $k_{r2} = 3000$, the vacuum pressure at the boundary forms instantaneously like $k_{r2} = +\infty$, so it is used below to represent the cases without time variation of vacuum pressure considered. Cases with various time-dependent factors are calculated, as shown in Fig. 3. It can be seen that the rate of consolidation decreases as the coefficient of time variation decreases, which is reasonable because smaller k_{r2} and k_{z2} means a slower increase of vacuum pressure at the boundaries, reducing the hydraulic gradient in the soil. Compared to the cases without time variation considered, when k_{r2} and k_{z2} are larger than 0.2, the influence of time variation only shows in the early stage of the consolidation, and the differences between cases with different k_{r2} and k_{z2} are not evident. In contrast, when $k_{r2} = k_{z2} = 0.01$, the consolidation rate is reduced a lot, and the influence exists during the whole consolidation process.

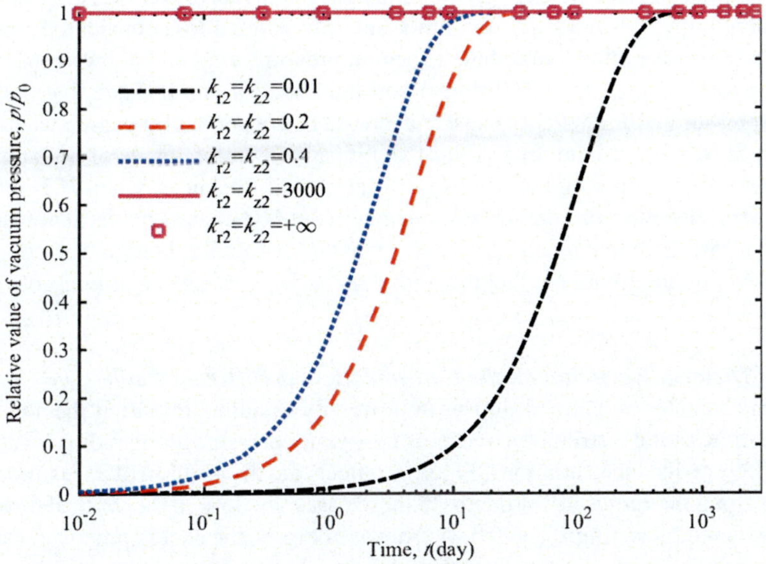

Fig. 2. Variation of vacuum pressure

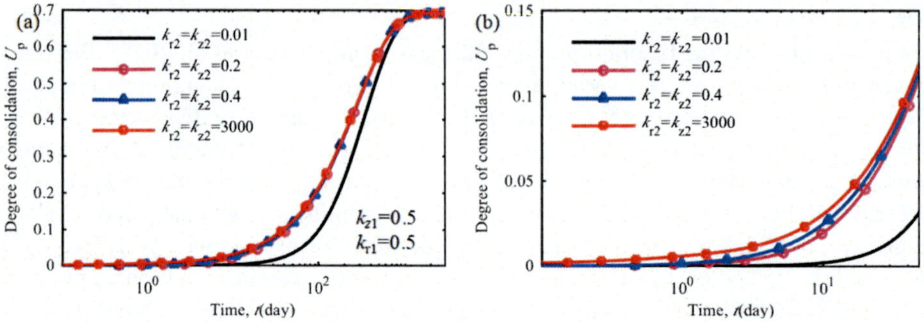

Fig. 3. Consolidation of cases with various k_{r2} and k_{z2} in (a) 3000 days; (b) 50 days

Effect of Distance Variation on the Average Consolidation Degree of Whole Soil Element

In this part, we focused on the influence of distance variation on the consolidation behavior. According to the common range of k_{r2} and k_{z2}, they are valued 0.4 in the following cases. Four conditions are applied in the cases to compare the effect of vertical loss and radial loss of vacuum pressure. $k_{r1} = k_{z1} = 0.5$ represents that both radial loss and vertical loss are considered, $k_{r1} = 0.5$ and $k_{z1} = 1$ represents that only radial loss is considered, $k_{r1} = 1$ and $k_{z1} = 0.5$ represent only vertical loss considered, and $k_{r1} = k_{z1} = 1$ represents uniform distributed vacuum pressure. Figure 4 shows consolidation under the four conditions with different soil thickness. It can be seen that the loss of vacuum pressure will inevitably cause a decrease of consolidation degree whatever the direction of loss is, and the drop is most significant when the losses of both directions are considered. Comparing the results of different soil thickness, when H is over r_e, a more evident decrease of consolidation degree would be caused by vertical loss rather than radial loss. Still, the results are inverse when H is smaller than r_e. The comparison results are consistent with the fact that radial seepage plays a more critical role in thick soil layers.

Effect of Distance Variation on the Consolidation at Different Radii

The main character of dredged slurries improved by vacuum pressure is the nonuniform consolidation, so the distribution of excess pore water pressure along radius is calculated under different distance variations. Figure 5 reflects the distribution of excess pore water pressure along the radius at 1/4 depth. In these cases, the loss in the vertical direction is also considered by setting $k_{z1} = 0.5$. As shown in Fig. 5, the excess pore water pressure in the clogging zone dissipates faster than the outside in the early stage of improvement, which is caused by the nonuniform consolidation. The hydraulic gradient is large in the clogging zone because this zone is adjacent to the drain. However, due to the low permeability in the clogging zone, the vacuum pressure penetrates slowly to the normal zone. Hence, the soil in that zone consolidates slowly at the initial time. Differences between cases with various k_{r1} are not evident initially but increase with consolidation processing. The ultimate distribution of excess pore water pressure is consistent with the distribution of vacuum pressure at the boundary. For cases with evident loss in the radial direction, the ultimate excess pore water pressure at different radii varies a lot.

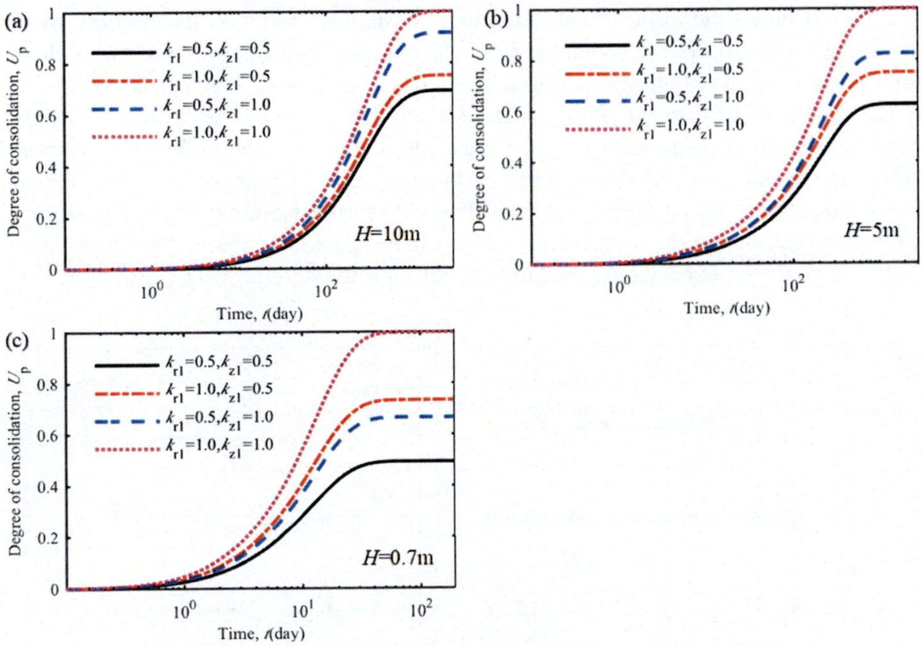

Fig. 4. Consolidation of cases with different distance variation

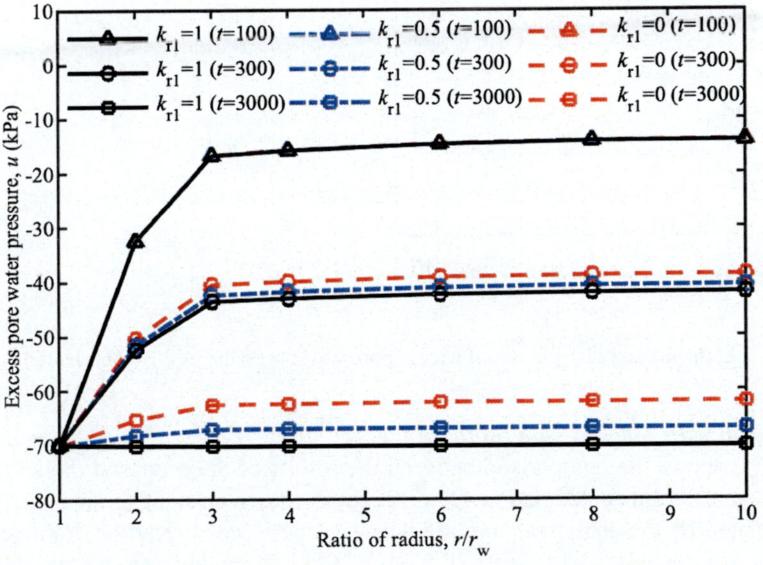

Fig. 5. Radial distribution of excess pore water pressure under difference radial loss

Figure 6 shows the influence of radial loss on the distribution of excess pore water pressure at different depth. It can be seen that the excess pore water pressure varies along radius at $H = 1/4$ and 3/4 from the start to the end. The difference between $k_{r1} = 0.5$ and $k_{r1} = 1$ can be investigated at 1/4 depth and 3/4 depth, but it cannot be seen at 1/2 depth. These facts show that the influence of vacuum loss along the radial direction is more evident at the position closer to the pervious boundaries. In these cases, both the bottom and top boundaries are pervious, and the 1/2 depth is farthest among the three positions, so the investigations in Fig. 6 is reasonable. It can be speculated that in the cases where the only top surface is pervious, the influence of k_{r1} will decrease with depth.

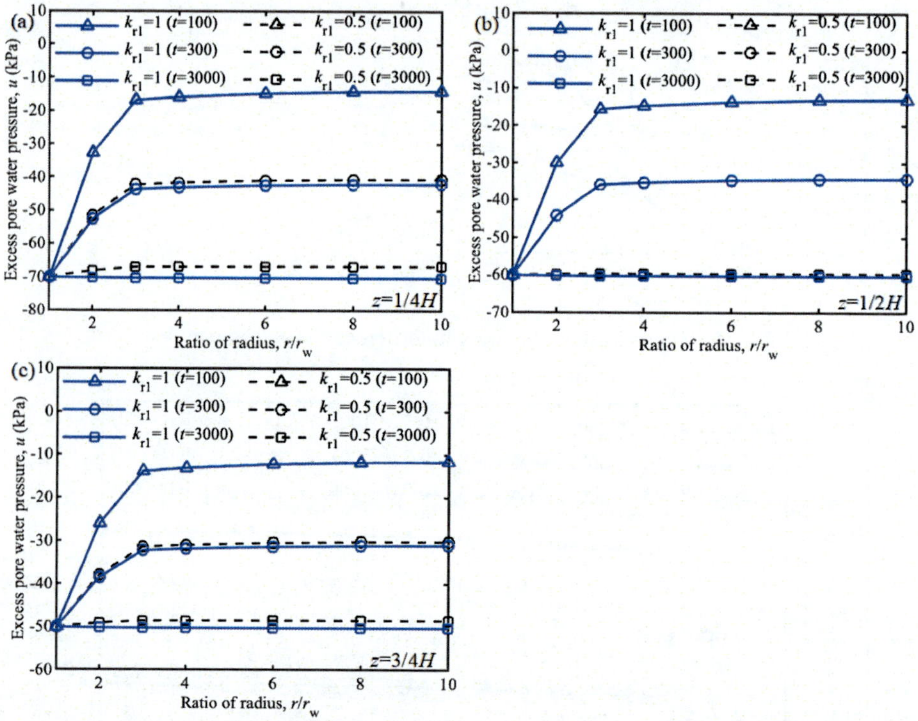

Fig. 6. Radial distribution of excess pore water pressure at different depths

Comparison with Small Strain Solution

Figure 7(a) shows the comparison between the results of the proposed model and that based on small strain consolidation. Here, the results based on small strain consolidation are calculated by the analytical solution where vacuum pressure at the drain generates gradually and decreases with depth (Liu et al. 2019). It can be seen that the difference between cases with various k_{z2} is larger in the results of small-strain consolidation solution than those of large-strain consolidation. As shown in Fig. 7(b), the consolidation degrees calculated by small strain solution are usually first larger and then lower than that calculated by large strain consolidation. Still, the difference between them varies with different k_{z2}. When k_{z2} is quite small, such as $k_{z2} = 0.01$, the difference mainly appears in the early stage. In contrast, when k_{z2} is more significant than 0.2, the difference is more evident in the middle stage.

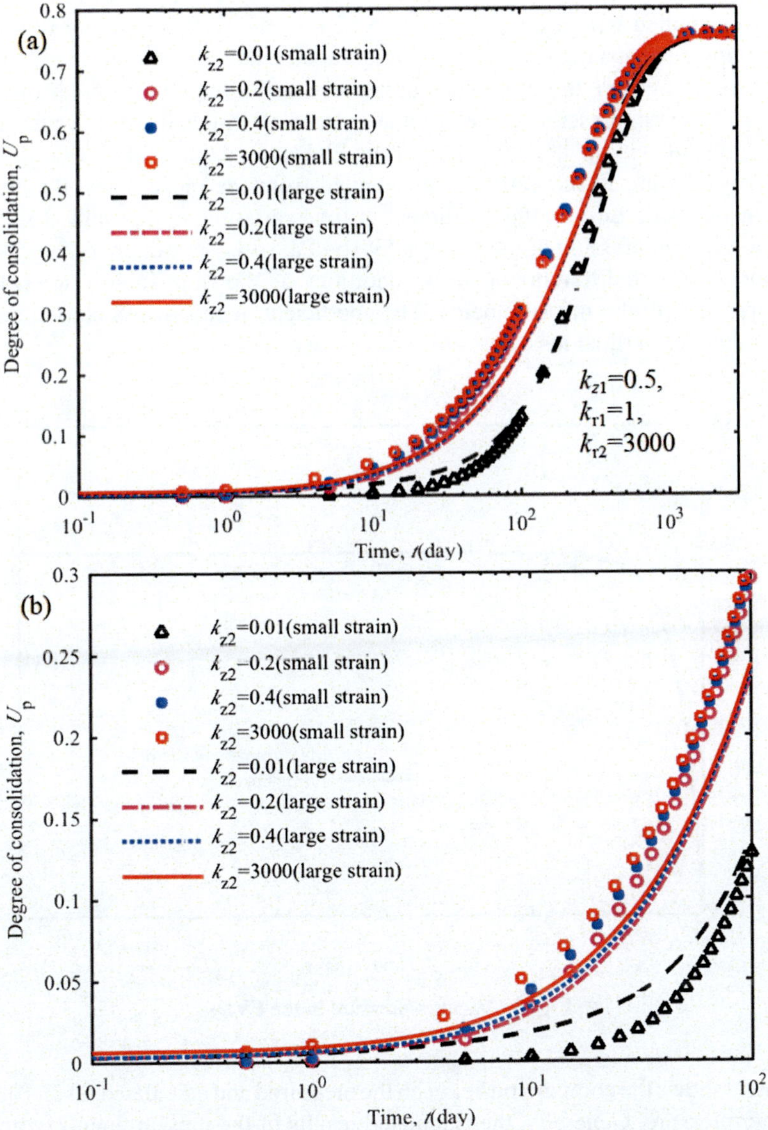

Fig. 7. Comparison between small strain solution and large strain solution in (a) 3000 days; (b) 100 days

Comparison with Field Tests

Measurements used in this part was obtained from Taizhou land reclamation project, which is located in Zhejiang Province, east of China. The thickness of dredged fills at this site was about 7 m. To improve the strength of the dredged fills, vacuum preloading method was applied with a magnitude of 85 kPa. In the wetland area of this site, the PVDs was installed at a distance of 0.9 m in the square pattern. Monitoring instruments were laid out to observe the improvement process, including a piezometer installed to measure the vacuum pressure in the drain at a depth of 3 m, and a pore water pressure transducer installed in the heart position of the PVDs at a depth of 4 m. Figure 8 shows the vacuum pressure in the early stage obtained from the piezometer. By fitting the measurements using Eq. (2), the coefficient of time variation, k_{z2}, is valued 1.752. The maximum vacuum pressure at the point is 59.16 kPa, so k_{z1} is valued to be 0.39 in this case. Four cases with different boundary conditions, as shown in Table 1, are calculated and compared with the measurements. The coefficient of radial loss is assumed to be equal with that of vertical loss.

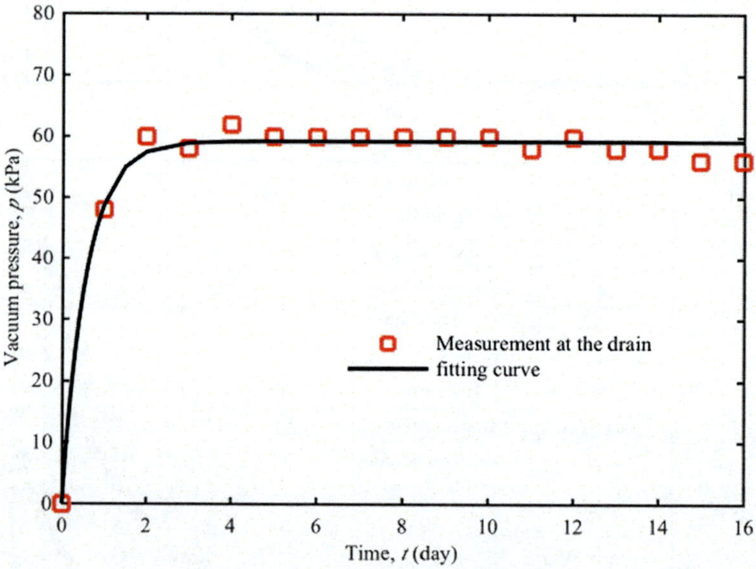

Fig. 8. Vacuum pressure in the PVDs

Figure 9 shows the comparison between the measured and calculated results of excess pore water pressure. Generally, the calculated results fit the measurements better when the vertical loss of vacuum pressure is considered. The case with both radial loss and vertical loss considered is closest to the measurements, where the radial loss slows down the dissipation of excess pore water pressure. The coefficient of time variation is large enough, so the impact of it can be ignored, as shown in Fig. 9.

Table 1. Value of variation coefficients

No.	Case A	Case B	Case C	Case D
G_s	2.73			
e_0	1.38			
C_c	0.39			
C_k	0.56			
σ_0' (kPa)	5			
k_{h0} (10^{-4} m/day)	5.18			
k_{v0} (10^{-4} m/day)	5.72			
k_{r1}	1	0.39	0.39	1
k_{r2} (day^{-1})	1.752	1.752	3000	1.752
k_{z1}	0.39	0.39	0.39	1
k_{z2} (day^{-1})	1.752	1.752	3000	1.752

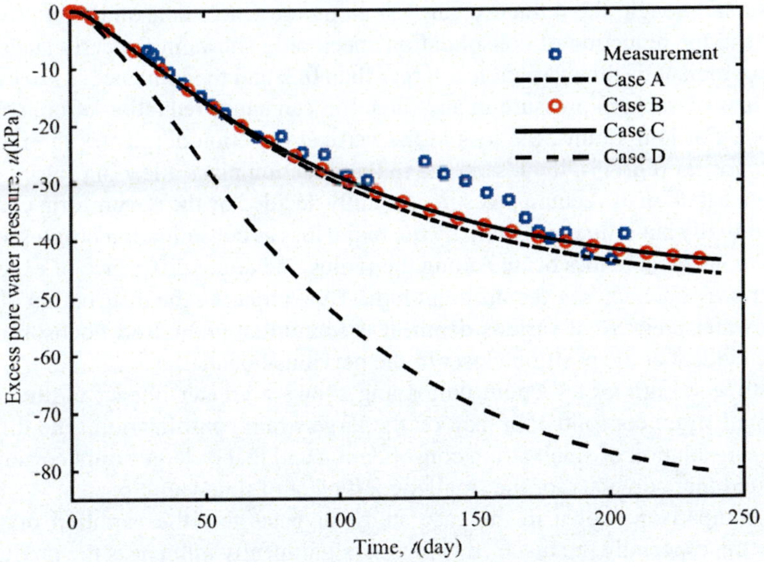

Fig. 9. Dissipation of excess pore water pressure

4 Limitations

Since this work is mainly focused on the effect of clogging on the primary consolidation behavior, some other changes occur in the soils are ignored, which are detailed below.

1. According to Fu et al. (2017), the improvement of dredged material consists of four parts, including the removal of pore water, reduction of the soil voids, cementation

of the soil particles, and electrolysis and evaporation of pore water. In this paper, there is only the improvement by the consolidation through the removal of water and reduction of void ratio considered, so the predicting strength may be underestimated.
2. Except for the primary settlement of consolidation, secondary settlement from creep can be a significant part in the deformation of soft soils (Hu et al. 2014), so it is better to consider creep of soil together with the large strain assumption to predict the settlement more accurately (Nguyen et al. 2020).

5 Conclusion

This paper builds a large strain consolidation model with the variation of vacuum pressure considered to calculate the consolidation of dredged slurries. The expressions of vacuum pressure at the vertical and radial drainage path are offered. The influence of time variation and distance variation, especially the radial loss along the radius, are studied by using FE method. The key findings are as follows:

1. The existence of time variation can reduce the rate of consolidation, and the influence increases with the decrease of k_{r2} and k_{z2}. When the coefficients of time variation are small enough, the influence can last along the whole consolidation process. In contrast, the reduction of consolidation speed only shows in the early stages when the coefficient of time variation is larger than 0.2, and the influence is not evident.
2. The loss of vacuum pressure in any direction can cause reduction of consolidation degree. For long drains, the loss in the vertical direction dominates the reduction, while for short drains, the loss in the radial direction plays a leading role.
3. The distribution of vacuum pressure is mainly decided by the nonuniform consolidation initially but influenced more by the radial loss as consolidation progressing. The more considerable loss occurs along the radius, the larger difference of excess pore water pressure shows at the different radii. By comparing the distribution of excess pore water pressure at various depth, it is found that nonuniform consolidation is more evident at the position closer to the pervious boundaries.
4. The time variation of vacuum preloading shows a greater impact on the solution of small strain consolidation than on the large strain consolidation. The difference between solution of small strain consolidation and that of large strain consolidation is found more obvious for the smaller coefficient of time variation.
5. The comparison shown in the case study suggests that the variation of vacuum pressure, especially the distance variation, significantly influences the prediction of large strain consolidation of dredged fills. For the case in Taizhou, better results is obtained considering both vertical and radial variation along with distance to the vacuum connector.

Acknowledgement. This work is supported by the National Key R&D Program of China (Grant No. 2017YFE0119500, Grant No. 2016YFC0800200), the Projects of International Cooperation and Exchanges NSFC (Grant No. 51620105008), the program of China Scholarships Council (No. 201906320247), the National Natural Science Foundation of China (Grant No. 51678319, Grant No. 51978533), the Natural Science Foundation of Shandong Province (Grant

No. ZR2016EEM40) and it received funding from the European Union's Horizon 2020 research and innovation programme Marie Skłodowska-Curie Actions Research and Innovation Staff Exchange (RISE) under grant agreement No. 778360.

References

Bao, S.F., Lou, Y., Dong, Z.L., Mo, H.H., Chen, P.S., Zhou, R.B.: Causes and countermeasures for vacuum consolidation failure of newly-dredged mud foundation. Chin. J. Geotech. Eng. **36**(7), 1350–1359 (2014a). (in Chinese). https://doi.org/10.11779/CJGE201407020

Bao, S., Mo, H., Dong, Z., Chen, P., Qiu, Q.C.: Research on transfer properties and distribution model of negative pressure in fresh hydraulic reclamation muck foundation. Rock Soil Mech. **35**, 3570–3576 (2014b)

Cai, Y., Qiao, H., Wang, J., Geng, X., Wang, P., Cai, Y.: Experimental tests on effect of deformed prefabricated vertical drains in dredged soil on consolidation via vacuum preloading. Eng. Geol. **222**, 10–19 (2017)

Chai, J., Hong, Z., Shen, S.: Vacuum-drain consolidation induced pressure distribution and ground deformation. Geotext. Geomembr. **28**(6), 525–535 (2010). https://doi.org/10.1016/j.geotexmem.2010.01.003

Deng, Y., Liu, L., Cui, Y.J., Feng, Q., Chen, X., He, N.: Colloid effect on clogging mechanism of hydraulic reclamation mud improved by vacuum preloading. Can. Geotech. J. **56**(5), 611–620 (2019). https://doi.org/10.1139/cgj-2017-0635

Geng, X., Yu, H.S.: A large-strain radial consolidation theory for soft clays improved by vertical drains. Géotechnique **67**(11), 1020–1028 (2017)

Gibson, R.E., Schiffman, R.L., Cargill, K.W.: The theory of one-dimensional consolidation of saturated clays. 2. Finite non-linear consolidation of thick homogeneous layers. Can. Geotech. J. **18**(2), 280–293 (1981)

Fang, Y., Guo, L., Huang, J.: Mechanism test on inhomogeneity of dredged fill during vacuum preloading consolidation. Mar. Georesour. Geotechnol. **37**(8), 1007–1017 (2019)

Fu, H., Cai, Y., Wang, J., Wang, P.: Experimental study on the combined application of vacuum preloading–variable-spacing electro-osmosis to soft ground improvement. Geosynth. Int. **24**(1), 72–81 (2017)

Hansbo, S.: Consolidation of fine-grained soils by prefabricated drains. In: Proceedings of the 10th International Conference on Soil Mechanics and Foundation Engineering, Balkema, Rotterdam, vol. 3, pp. 677–682 (1981)

Hu, Y.Y., Zhou, W.H., Cai, Y.Q.: Large-strain elastic viscoplastic consolidation analysis of very soft clay layers with vertical drains under preloading. Can. Geotech. J. **51**(2), 144–157 (2014)

Indraratna, B., Zhong, R., Fox, P.J., Rujikiatkamjorn, C.: Large-strain vacuum-assisted consolidation with non-Darcian radial flow incorporating varying permeability and compressibility. J. Geotech. Geoenviron. Eng. **143**(1), 04016088 (2017)

Lei, H., Lu, H., Liu, J., Zheng, G.: Experimental study of the clogging of dredger fills under vacuum preloading. Int. J. Geomech. **17**(12), 04017117 (2017). https://doi.org/10.1061/(ASCE)GM.1943-5622.0001028

Liu, J., Lei, H., Zheng, G., Zhou, H., Zhang, X.: Laboratory model study of newly deposited dredger fills using improved multiple-vacuum preloading technique. J. Rock Mech. Geotech. Eng. **9**(5), 924–935 (2017). https://doi.org/10.1016/j.jrmge.2017.03.003

Liu, S.J., Geng, X.Y., Sun, H.L., Cai, Y.Q., Pan, X.D., Shi, L.: Nonlinear consolidation of vertical drains with coupled radial-vertical flow considering time and depth dependent vacuum pressure. Int. J. Numer. Anal. Meth. Geomech. **43**(4), 767–780 (2019)

Nguyen, B.P., Do, T.H., Kim, Y.T.: Large-strain analysis of vertical drain-improved soft deposit consolidation considering smear zone, well resistance, and creep effects. Comput. Geotech. **123**, 103602 (2020)

Pu, H., Yang, P., Lu, M., Zhou, Y., Chen, J.N.: Piecewise-linear large-strain model for radial consolidation with non-Darcian flow and general constitutive relationships. Comput. Geotech. **118**, 103327 (2020)

Tang, T.Z., Huang, J.Q., Guan, Y.F., Chen, H.B., Cheng, W.Z.: Experimental study on dredged fill sludge improved by vacuum preloading. . Port Waterway Eng. **440**, 115–122 (2010). https://doi.org/10.3969/j.issn.1002-4972.2010.04.027. (in Chinese)

Zeng, L.L., Hong, Z.S., Cui, Y.J.: Time-dependent compression behaviour of dredged clays at high water contents in China. Appl. Clay Sci. **123**, 320–328 (2016). https://doi.org/10.1016/j.clay.2016.01.039

Zhou, Y., Chai, J.C.: Equivalent 'smear' effect due to non-uniform consolidation surrounding a PVD. Géotechnique **67**(5), 410–419 (2017). https://doi.org/10.1680/jgeot.16.P.087

Preliminary Investigation on Overall Permeability of Granular Mixed Materials

Kaiqi Li[1,2], Dianqing Li[1,2], Kang Yan[1,2](✉), Chia Zarei[1,2], and Yong Liu[1,2]

[1] State Key Laboratory of Water Resources and Hydropower Engineering Science, Institute of Engineering Risk and Disaster Prevention, Wuhan University, Wuhan 430072, People's Republic of China
{kq_jane,dianqing,yk,chia.zarei,liuy203}@whu.edu.cn

[2] Key Laboratory of Rock Mechanics in Hydraulic Structural Engineering of the Ministry of Education, Wuhan University, Wuhan 430072, People's Republic of China

Abstract. Many geomaterials are mixed composites and the permeability coefficient of each phase differs widely. Determining the overall permeability coefficient of mixed materials is still a challenging task due to the heterogeneity. Few studies have focused on the physical mechanism of the hydraulic property of mixed geomaterial. To address this lack of knowledge, a constant head permeability test and numerical simulation based on CFD (computational fluid dynamics)-DEM (discrete element method) are conducted to investigate seepage performance of mixture specimens. To explore the effects of content of each phase on the overall permeability coefficient, glass beads as inclusion phase are distributed uniformly and randomly in quartz sand, and different contents of inclusions ranging from 0% to 60% are considered. The experimental and numerical results show that the overall permeability coefficient increases with the content of the glass beads and porosity. It can be seen that the overall permeability coefficient is not only related to the content and property of each component, but also more determined by the interconnected paths. This work provides new sights for the future research of hydraulic property of mixed materials.

Keywords: Permeability coefficient · Mixed material · Constant head test · Numerical method · Porosity

1 Introduction

Seepage properties are the most important features of geomaterials that allow fluids to flow through the interconnected voids, which can be quantified by the permeability coefficient [1, 2]. Pervious materials have attracted strong attentions due to their application in the geological field, including slope stability analysis [3, 4], deformation of subgrade [5], the dam seepage [6], levee seepage problem [7], shale reservoirs [8], hazardous waste landfills [9] and so on. In general, mixed materials are generally composed of different components varied in the permeability coefficient. Terzaghi and Peck (1967) indicated that the value of hydraulic conductivity can vary over a range by several

B. Fatahi et al. (Eds.): GeoChina 2021, SUCI, pp. 97–109, 2021.
https://doi.org/10.1007/978-3-030-79854-3_8

orders of magnitude, which is highly dependent upon particle size [10]. The permeability coefficient of mixed materials is more complex due to the heterogeneity. However, the permeability characteristics of mixtures are mostly regarded as homogeneous materials in the process of design and construction practical projects. Therefore, there is still a lack of clear understanding of how the internal structure affects the overall permeability coefficient of mixed material, which is a major stumbling block to the future research and engineering application of composite geomaterial. It is therefore necessary to figure out the relationship between the internal structure and the overall hydraulic property.

A great number of studies reported on the estimation of the permeability coefficient, which can be classified into direct and indirect methods. Direct methods are primarily test methods including laboratory experimental tests [11–15] and field tests [16], which tend to be time-consuming and labor-intensive. Some scholars investigated the permeability using indirect methods like fractal model [17–20] and computed tomography (CT) technique [3, 21, 22], whereas few of them revealed the hydraulic mechanism of fluid flow. Other alternative methods are to empirically deduce the coefficient of permeability from the results of other tests such as the specified matric suction values [23–25] or empirical formulas, like the Terzaghi formula [26], Hazen formula [27] and Kozeny-Carman formula [28]. Nevertheless, those empirical formulas for the penetration properties are not suitable for mixed materials as their stronger heterogeneity and wide difference in each phase. In addition, a large amount of literature has studied the permeability coefficient of geomaterials, but those ignored the mixed characteristics and the effect of each phase on the overall performance. A preliminary study for measuring the permeability coefficient of mixed materials is useful to understand the seepage properties of mixed materials, attracting increasing attention to the overall permeability coefficient of mixed materials.

As each phase may greatly differ in the permeability coefficient, a constant head permeameter was utilized to measure the overall permeability coefficient of mixture materials where glass beads were blended into the quartz sand. To explore the influence of porosity and each phase on the overall permeability coefficient of mixtures, a numerical model based on CFD-DEM was adopted to calculate the permeability coefficient (k). The experimental and numerical simulation results were compared to the traditional models for illustrating the relationship between macro hydraulic response and internal structure, which are beneficial to understand the hydraulic properties of mixed materials.

2 Experimental Method

2.1 Testing Equipment

The constant head permeameter is often used for measuring the permeability coefficient of coarse-grained soil, while variable head permeameter is more suitable for fine-grained soil. Herein, seepage properties of mixtures were measured by TST-70 constant head permeameter (GB/T50123-1999 Standard for soil test method) in this work. Given in Fig. 1(a), the testing system for measuring the permeability coefficient consists of four parts: (I) water supply system was mainly used for providing the sufficient water during the whole test; (II) hydraulic pressure system can display the water head difference between two pressure tubes; (III) sample system was the main parts of permeameter containing the prepared samples; and (IV) measure system was utilized to measuring

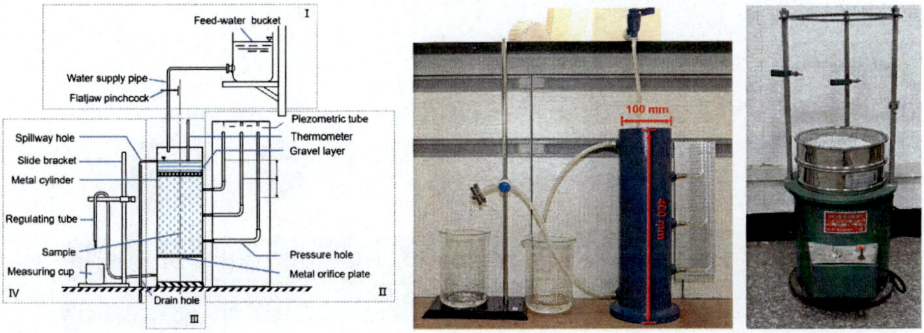

(a) (b) (c)

Fig. 1. (a) Schematic diagram of permeameter; (b) Constant head permeameter (TST-70); (c) Electric vibrating screen.

the amount of water that flowed out of the permeameter within the required time. As presented in Fig. 1(b), the inner diameter and height of the permeameter are 100 mm and 400 mm, respectively. The electric vibrating screen was used for screening particles (Fig. 1(c)).

2.2 Sample Preparation

For exploring the seepage properties of mixed materials, glass beads and quartz sands were mixed together as specimens. In order to control the particle size effect, the maximum grain size of samples could not exceed 1/5 of the internal diameter of the permeameter [3]. Therefore, the maximum particle size was set at 20 mm as shown in Table 1. Referring to Fig. 2, glass beads were regarded as inclusion phase with different diameters varying from 5 mm to 20 mm, which are made by glass and strong enough to resist the forces of vibration and compaction during the test procedure. Besides, the spherical shape of glass beads also makes a positive contribution during the movement of water through the mixed sample since the smooth curved surface can reduce resistance to water flow. The matrix phase employed in the experiment was the quartz sand with irregular shape and the size is less than 5 mm (see Fig. 3). To gain insight into the influence of each component on the overall permeability of mixed materials, a series of specimens with different volume fraction (0%–60%) of glass beads were prepared. The mixture consisting of quartz sand and glass beads was shown in Fig. 4, where the volume fraction of glass beads with different particle sizes are uniformly and randomly distributed in the matrix part.

Table 1. Properties of particles in the mixed sample.

Material	Phase	Particle size (mm)
Quartz sand	Matrix phase	1–5
Glass beads	Inclusion phase	5, 6, 7, 8, 10, 12, 16, 18, 20

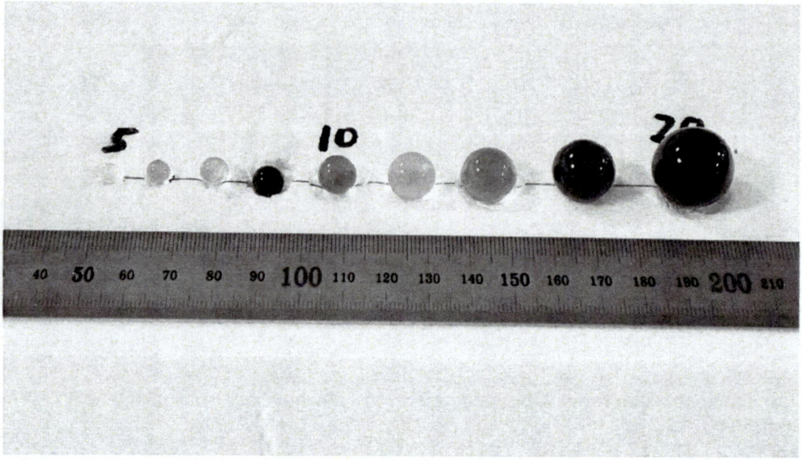

Fig. 2. Glass beads with different sizes (from 5 mm to 20 mm).

(a) (b) (c)

Fig. 3. Quartz sand (a) size = 1 mm; (b) size = 5 mm; (c) quartz sand mixture with different sizes ranging from 1 mm to 5 mm.

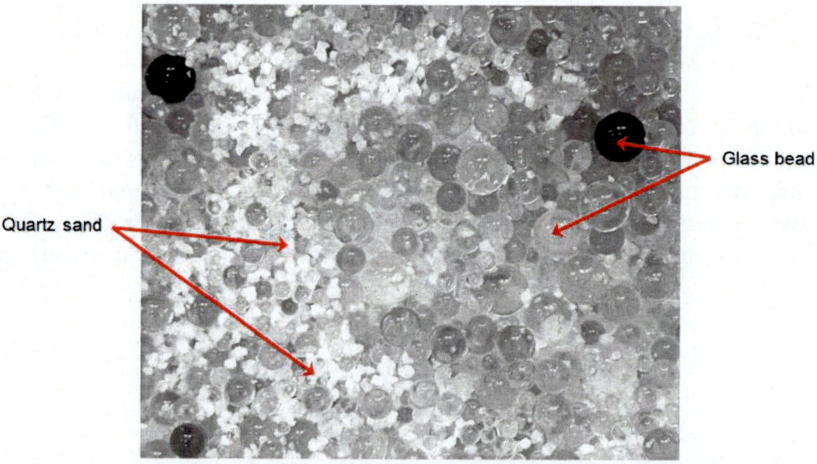

Fig. 4. A prepared mixture sample where glass beads were blended into quartz sand.

2.3 Testing Procedures

The prepared mixture (Fig. 4) was subsequently put into the permeameter layer by layer, and the thickness of each layer was about 2–3 cm. After that, the water supply pipe valve was kept open until the whole specimen was filled with water, and then each specimen was saturated for 24 h before conducting the test. In addition, the water used in the experiment should be degassed and kept the temperature 3–4° higher than room temperature to prevent the water decomposing into some bubbles when entering the sample. During the experiment, the amount of water moving through the saturated sample is collected over a period of time. After changing the water head, the same procedure was conducted when water levels become stable. Every sample was measured repeatedly six times for reducing the errors caused by experimental measurement. Moreover, the average values of the overall permeability coefficient from six tests were calculated and recorded. To explore the effect of each component on the overall permeability coefficient of mixtures, a series of constant head permeability tests are carried out by varying the content of glass beads.

2.4 Calculation Theory

A constant head test for calculating the permeability coefficient is based on Darcy's law that describes the water flow model in porous materials and established the relationship between the water flow amount and permeability coefficient. According to the relationship, the seepage quantity through the sample in per unit time is proportional to the permeability coefficient. The specific expression is shown as follows [29].

$$Q/t = k\frac{\Delta H}{L}\mathrm{A} = ki\mathrm{A} \tag{1}$$

where Q is the amount of water flow out from the permeameter within t seconds; k is the permeability coefficient of the specimen; i is the hydraulic gradient, equals $\Delta H/L$; L is the distance between adjacent pressure tubes; ΔH is the average value of hydraulic pressure difference between pressure tubes; and A is the cross-sectional area of the sample.

3 Numerical Method

With the improvement of computing efficiency, numerical simulation has become a more and more important research method. Compared with the test method, numerical simulation is a time saving and efficient way to explore the effect of a single variable on the results. In this work, the numerical simulation is conducted based on CFD-DEM method, which is applicable to simulate the interaction between granular materials and fluids [30–32]. The numerical samples are generated using the three-dimensional discrete element method code PFC^{3d}. During the simulation process, the flow field is calculated under the precondition of incompressible flow and satisfies Eqs. (2) and (3).

$$\vec{v} = -\frac{k}{\mu\varepsilon}\vec{\nabla} p \tag{2}$$

$$\vec{\nabla} \cdot \vec{v} = 0 \tag{3}$$

where $\vec{v}$ is the fluid velocity; k is the permeability; μ is the fluid viscosity; ε is the porosity; p is the fluid pressure; and $\vec{\nabla}$ is Hamiltonian operator.

Combining Eq. (2) and Eq. (3) yields the governing equation for flow field calculation.

$$\vec{v} \cdot (\frac{\mathrm{k}}{\mu\varepsilon}\vec{\nabla} p) = 0 \tag{4}$$

To solve this governing equation numerically, the values of μ, ε and k on each discrete fluid element was determined as follows. The value of μ is set as 1×10^{-3} Pa·s, namely the dynamic viscosity of water. The value of ε is the porosity of the fluid element,

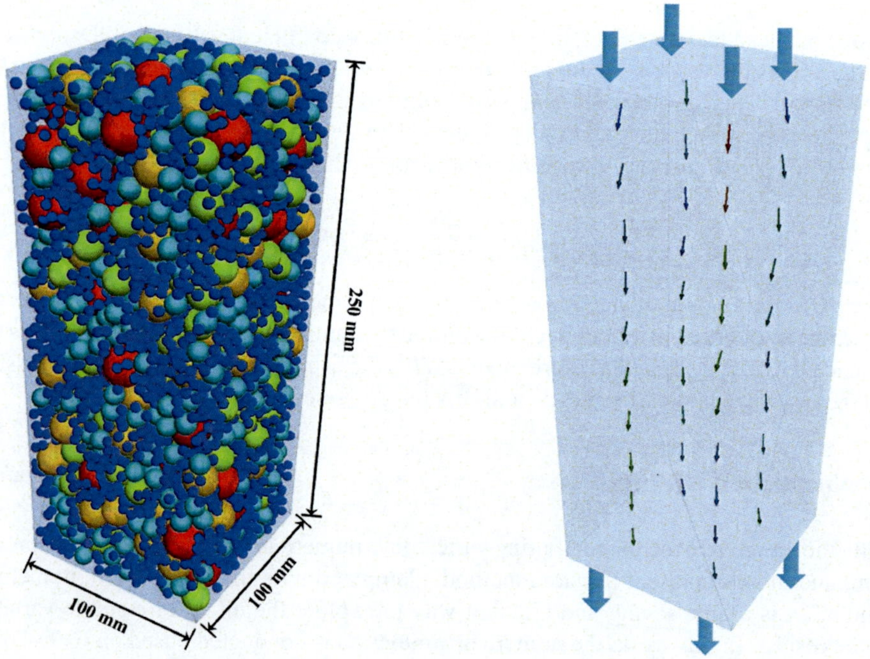

(a) (b)

Fig. 5. (a) Numerical model and model size; the content of glass beads of the granular column is 60%; the dark blue particles are quartz sand and others color particles are glass beads; (b) Boundary condition and flow field distribution; the top and bottom surfaces of the model are subjected to fixed water pressure, while the other four surfaces are impervious.

namely the percentage of residual volume of the element after removing particles. k can be determined based on the Kozeny-Carman relation [33].

$$k(\varepsilon) = \begin{cases} \frac{1}{180}\frac{\varepsilon^3}{(1-\varepsilon)^2}(2r)^2 & \varepsilon \le 0.7 \\ k(0.7) & \varepsilon > 0.7 \end{cases} \quad (5)$$

where r is the mean radius of PFC particles. Both ε and k are calculated automatically in the PFC^{3d}.

Figure 5(a) illustrates a granular column sample with 60% content of glass beads (f_g); the specimen size of 100 × 100 × 250 mm is utilized for the computation efficiency. Quartz sand is simulated by balls with radius of 1.5 mm, and the radii of glass beads are range from 2.5 mm to 10 mm. The proportion of glass beads with different particle sizes remains the same and then randomly generated in a cuboid container. The position of each particle is fixed once the simulated model is generated. Besides, the water head difference between the top and the bottom surface is 0.7 kPa and other surfaces of the sample are impervious. The boundary condition and flow velocity distribution are shown in Fig. 5(b). In order to investigate the influence of inclusion content and porosity (n), various samples with different parameters are considered. Herein, four typical porosity values are selected; those are 0.25, 0.30, 0.35 and 0.40. The content of glass beads, same as the experiment, varies from 0% to 60%.

4 Results and Discussions

4.1 Traditional Model Analysis

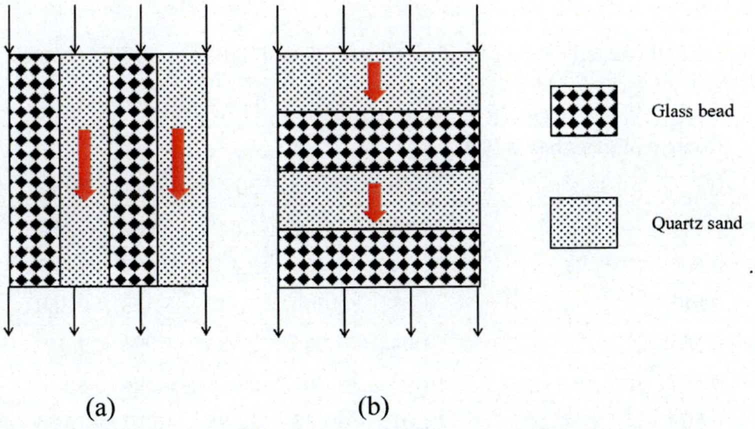

Fig. 6. Prediction models (a) parallel model (b) series model; the arrows represent are the water flow direction.

There are several classical models for estimating the permeability coefficients of mixed materials. Series and parallel models are often adopted to simplify the analysis of heterogeneous materials, especially for two-phase material, which provides the upper and lower bounds of the overall properties of mixed materials [34–36]. The schematic diagrams of series and parallel models are shown in Fig. 6. Under the head pressure, water tends to bypass the impervious medium and flow through the materials with larger permeability coefficients. Glass beads are regarded as impervious material without any internally connected pores, while the quartz sand is more permeable and there are a lot of voids between adjacent sand due to its irregular shape. Therefore, water mostly passes through the quartz sand because the glass beads are impervious in the parallel model (Fig. 6(a)), however, the flow of water is impeded by the impervious glass beads in the series arrangement (Fig. 6(b)). Thus, one might expect that the increasing content of glass beads contributes to the decrease of the overall permeability of mixed materials regardless of whether in parallel or series arrangement. However, the results obtained from the constant head permeability test and numerical simulation model show an opposite curve trend.

4.2 Experimental Results

The overall permeability coefficients of the mixed specimens measured by permeameter with different contents of glass beads were shown in Table 2 and Fig. 7. The results revealed that the permeability coefficient of the mixture sample without glass beads is 5.07×10^{-5} m/s. An interesting observation is that the permeability coefficient significantly increased with the increment of the content of inclusion, and the rate of increase gradually rose, particularly when the glass beads content exceeded 30%. This changing trend wasn't consistent with previous analysis based on the series or parallel model and showed an opposite result, which may be due to the fact that the internal structure of the mixed samples had changed.

Table 2. Permeability coefficient of mixture samples with different glass beads content (Unit: $\times 10^{-4}$ m/s)

Test no.	Content of glass beads (%)						
	0%	10%	20%	30%	40%	50%	60%
1	0.513	1.157	1.492	2.645	5.728	9.635	11.831
2	0.506	1.145	1.488	2.477	5.699	9.442	12.795
3	0.517	1.081	1.359	2.438	5.578	9.369	12.335
4	0.505	1.042	1.345	2.493	5.096	9.029	10.253
5	0.504	1.031	1.267	2.418	4.394	8.834	10.747
6	0.502	1.015	1.235	2.295	4.297	8.993	11.283
Average	0.508	1.079	1.364	2.461	5.132	9.217	11.541

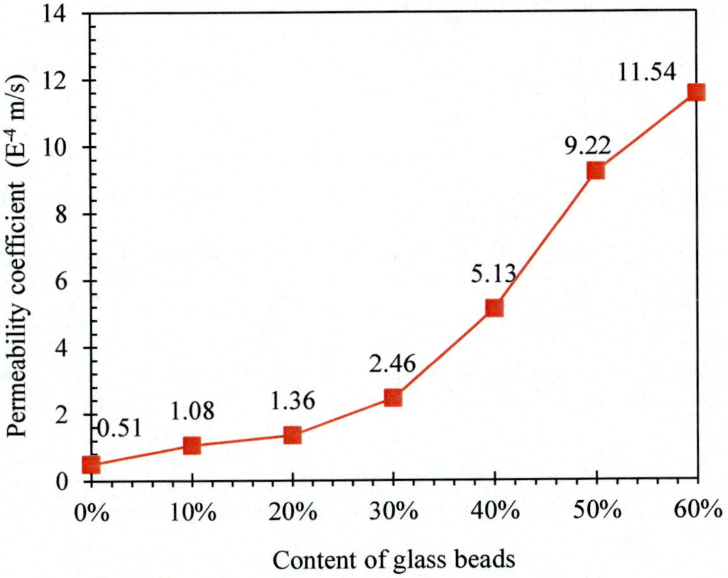

Fig. 7. Overall permeability coefficient of mixture with different contents of glass beads

Glass beads were likely to be separated and the holes between glass beads were wrapped by quartz sand with small particle size when glass beads content is relatively lower. As the volume fraction of glass beads increases, spherical glass beads tend to contact each other. It's more difficult for quartz sand with small size to fill the hole between beads perfectly due to the arcuate surface of glass beads. As illustrated in Fig. 8, the increment of glass beads content of samples provides larger space and connected seepage paths for water to flow, thereby and the overall permeability coefficient of the mixture sample increased accordingly. Thus, the phenomenon of percolation can be attributed to the existence of interconnected paths. It means that the permeability coefficient of mixed materials depends not only on the volume fraction of each phase, but also on the interconnected paths. To figure out how the internal structure influences the permeability coefficient of mixture samples, the numerical results are illustrated in the following section.

4.3 Numerical Results

The numerical simulation for calculating the permeability coefficient was carried out on mixture samples with different glass beads content as well as porosity. The considered glass beads content during simulation procedure are 0%, 10%, 20%, 30%, 40%, 50% and 60%. To study how the existence of voids affects the permeability test, four typical models with different porosity are performed for each sample.

Figure 9(a) plots the numerical simulation results and indicates that the permeability coefficients of simulation samples are increasing with the increment of porosity. As glass beads content increases, the increasing trend of permeability coefficients is more

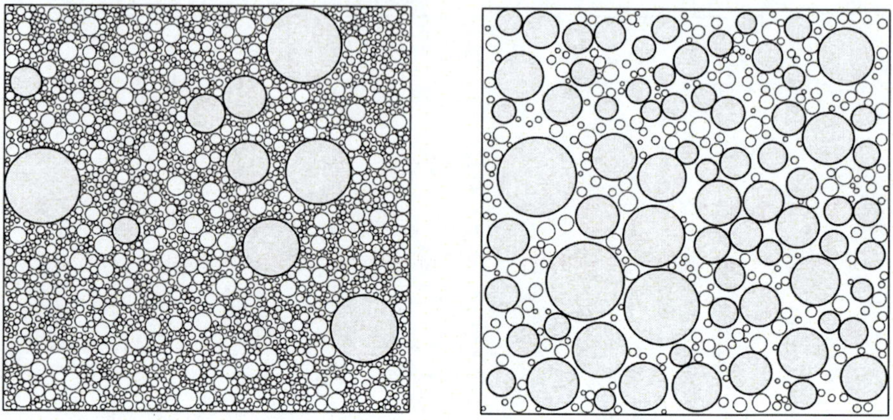

(a) (b)

Fig. 8. Schematic interpretations of mixture with different glass beads content; (a) $f_g = 12\%$; (b) $f_g = 48\%$. The circular particles with the bold black line are glass beads, and other particles represent quartz sand.

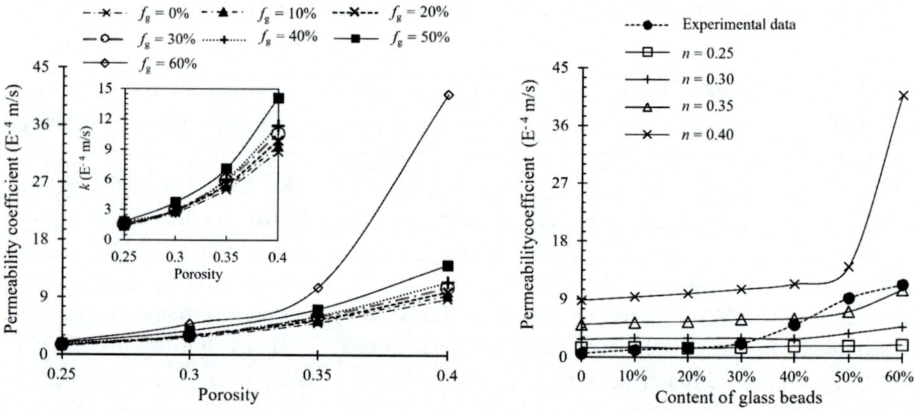

(a) (b)

Fig. 9. (a) The relationship between the permeability coefficient of numerical models with different glass beads content and porosity; (b) The relationship between the permeability coefficient of numerical models with different porosity and glass beads content.

obvious, especially when the glass beads content is over 50%. Therefore, the threshold of inclusion content is 50% which meaning that if a mixture has a value of f_g greater than 50%, the permeability coefficients of sample tend to an exponential growth. Given in Fig. 9(b), one can notice that the constant head test data fall within the range of numerical simulation results, which demonstrates the validity and effectiveness of the numerical model. Being consistent with the trend of constant head test results, higher content of glass beads promotes the improvement of permeability coefficient of samples that porosity is constant. However, the changing curve is not significant when the porosity is no more than 0.35.

5 Conclusions

Geomaterials play an important role in geotechnical engineering and the evaluation of their permeability coefficient is complicated due to their heterogeneity. The objective of this work was to estimate the permeability coefficient of the mixed geomaterials by considering the effect of internal structure, i.e., porosity and each component contents. In this regard, glass beads that represent the inclusion phase were mixed with quartz sand in the laboratory, and the prepared mixtures were tested by a constant head permeameter. A corresponding numerical model based on the CFD-DEM method was carried out, and the effects of the different contents of porosity and glass beads are considered. The permeability coefficient results in both numerical simulation and experiment indicate that the internal structure (f_g and n) had a remarkable influence on the overall permeability of mixture materials, the value of k significantly increased with the increment of the content of inclusion and porosity; the rate of increase gradually rose, particularly when the glass beads content exceeded 30% and the porosity exceeded 0.35. Especially, the threshold of inclusion content is 50%, and the permeability coefficient of mixture grows dramatically when the value of f_g exceeds the threshold level. The results from experiment and numerical simulation indicate that the variation trend is different from the analysis according to the traditional prediction models. The permeability coefficient of mixture materials is not only determined by the content of each phase, but also affected by the interconnected paths. The findings of the work provide a further understanding of the mechanism of how the internal structure affect the permeability coefficient of mixed materials.

Acknowledgements. This research is supported by the NRF-NSFC 3rd Joint Research Grant (Earth Science) (Grant No. 41861144022) and the National Natural Science foundation of China (Grant No. 51879203, 52079099).

References

1. Doyen, P.M.: Permeability, conductivity, and pore geometry of sandstone. J. Geophys. Res. Solid Earth **93**(B7), 7729–7740 (1988)
2. Moldrup, P., Olesen, T., Komatsu, T., Schjønning, P., Rolston, D.E.: Tortuosity, diffusivity, and permeability in the soil liquid and gaseous phases. Soil Sci. Soc. Am. J. **65**(3), 613–623 (2001)
3. Johari, A., Talebi, A.: Stochastic analysis of rainfall-induced slope instability and steady-state seepage flow using random finite-element method. Int. J. Geomech. **19**(8), 04019085 (2019)
4. Dong, H., Huang, R., Gao, Q.F.: Rainfall infiltration performance and its relation to mesoscopic structural properties of a gravelly soil slope. Eng. Geol. **230**, 1–10 (2017)
5. Zhang, J., Ma, G., Ming, R., Cui, X., Li, L., Xu, H.: Numerical study on seepage flow in pervious concrete based on 3D CT imaging. Constr. Build. Mater. **161**, 468–478 (2018)

6. Wu, Y., Zhang, B., Yu, Y., Zhang, Z.: Consolidation analysis of Nuozhadu high earth-rockfill dam based on the coupling of seepage and stress-deformation physical state. Int. J. Geomech. **16**(3), 04015085 (2016)
7. Wang, R.H., Li, D.Q., Wang, M.Y., Liu, Y.: Deterministic and probabilistic investigations on piping occurrence during tunnelling through spatially variable soils. ASCE-ASME J. Risk Uncertainty Eng. Syst. Part A Civ. Eng. **7**, 04021009 (2021)
8. Liu, J., Liang, X., Xue, Y., Fu, Y., Yao, K., Dou, F.K.: Investigation on crack initiation and propagation in hydraulic fracturing of bedded shale by hybrid phase-field modeling. Theoretical Appl. Fracture Mech. **108**, 102651 (2020)
9. Harrop-Williams, K.: Clay Liner permeability: evaluation and variation. J. Geotech. Eng. **111**(10), 1211–1225 (1985)
10. Terzaghi, K., Peck, R.B.: Soil Mechanics in Engineering Practice, p. 729. John Wiley and Sons, New York (1967)
11. Kenney, T., Lau, D., Ofoegbu, G.: Permeability of compacted granular materials. Can. Geotech. J. **21**(4), 726–729 (1984)
12. Illman, W.A., Hughson, D.L.: Stochastic simulations of steady state unsaturated flow in a three-layer, heterogeneous, dual continuum model of fractured rock. J. Hydrol. **307**(1–4), 17–37 (2005)
13. Ueng, T.S., Wang, Z.F., Chu, M.C., Ge, L.: Laboratory tests for permeability of sand during liquefaction. Soil Dyn. Earthq. Eng. **100**, 249–256 (2017)
14. Juang, C.H., Holtz, R.D.: Fabric, pore size distribution, and permeability of sandy soils. J. Geotech. Eng. **112**(9), 855–868 (1986)
15. Alam, J., Muzzammil, M., Singh, H.P., Gupta, P.: Permeability of stratified soils for flow normal to bedding plane. Aquatic Procedia **4**, 660–667 (2015)
16. Elhakim, A.F.: Estimation of soil permeability. Alex. Eng. J. **55**(3), 2631–2638 (2016)
17. Xu, Y.F., Sun, D.A.: A fractal model for soil pores and its application to determination of water permeability. Physica A Stat. Mech. Appl. **316**(1–4), 56–64 (2002)
18. Tan, X.H., Liu, J.Y., Li, X.P., Zhang, L.H., Cai, J.: A simulation method for permeability of porous media based on multiple fractal model. Int. J. Eng. Sci. **95**, 76–84 (2015)
19. Li, B., Liu, R., Jiang, Y.: A multiple fractal model for estimating permeability of dual-porosity media. J. Hydrol. **540**, 659–669 (2016)
20. Shen, X., Li, L., Cui, W., Feng, Y.: Improvement of fractal model for porosity and permeability in porous materials. Int. J. Heat Mass Transf. **121**, 1307–1315 (2018)
21. Jivkov, A.P., Hollis, C., Etiese, F., McDonald, S.A., Withers, P.J.: A novel architecture for pore network modelling with applications to permeability of porous media. J. Hydrol. **486**, 246–258 (2013)
22. Ramandi, H.L., Mostaghimi, P., Armstrong, R.T.: Digital rock analysis for accurate prediction of fractured media permeability. J. Hydrol. **554**, 817–826 (2017)
23. Zhai, Q., Rahardjo, H.: Estimation of permeability function from the soil–water characteristic curve. Eng. Geol. **199**, 148–156 (2015)
24. Huang, S.Y., Fredlund, D.G., Barbour, S.L.: Measurement of coefficient of permeability of unsaturated soil. In: Paper presented at the Proceedings of the First International Conference on Unsaturated Soils. AA Balkema, Rotterdam (1995)
25. Huang, S., Fredlund, D., Barbour, S.: Measurement of the coefficient of permeability for a deformable unsaturated soil using a triaxial permeameter. Can. Geotech. J. **35**(3), 426–432 (1998)
26. Terzaghi, K., Peck, R.B., Mesri, G.: Soil Mechanics in Engineering Practice. John Wiley & Sons, New York (1996)
27. Hazen, A.: Discussion: dams on sand foundations. Trans. Am. Soc. Civ. Eng. **73**(11), 199–203 (1911)

28. Carman, P.C.: Fluid flow through granular beds. Chem. Eng. Res. Des. **75**(1), 32–48 (1997)
29. Darcy, H.P.C.: Les Fontaines Publiques de la Ville de Dijon. Victor, Dalmont, Paris (1856)
30. Abdelhamid, Y., El Shamy, U.: Pore-scale modeling of fine-particle migration in granular filters. Int. J. Geomech. **16**(3), 04015086 (2016)
31. Shi, Z., Zheng, H., Yu, S., Peng, M., Jiang, T.: Application of CFD-DEM to investigate seepage characteristics of landslide dam materials. Comput. Geotech. **101**, 23–33 (2018)
32. Yan, C., Zheng, H.: Three-dimensional hydromechanical model of hydraulic fracturing with arbitrarily discrete fracture networks using finite-discrete element method. Int. J. Geomech. **17**(6), 04016133 (2017)
33. Carrier III, W.D.: Goodbye, Hazen; hello, Kozeny-Carman. J. Geotech. Geoenvironmental Eng. **40**(120), 1054–1056 (2003)
34. Li, K.Q., Li, D.Q., Li, P.T., Liu, Y.: Meso-mechanical investigations on the overall elastic properties of multi-phase construction materials using finite element method. Constr. Build. Mater. **228**, 116727 (2019)
35. Li, K.Q., Li, D.Q., Liu, Y.: Meso-scale investigations on the effective thermal conductivity of multi-phase materials using the finite element method. Int. J. Heat Mass Transfer **151**, 119383 (2020)
36. Jia, G.S., et al.: Review of effective thermal conductivity models of rock-soil for geothermal energy applications. Geothermics **77**, 1–11 (2019)

used void ratio, velocity fields, average pure rotation rates (APR), stress fields, velocity fields, contact force chains and bond breakage distributions to study strain localization showing the significance of combining micro and macro-mechanical characteristics.

Material instability related to the micro-structure and geometric instability related to boundary conditions are associated with deformation of soil and strain localization. In granular soil, rotation of grains and breakage of force chains as a result of reduction in contact between particles has been observed to be highest in particles inside the shear band as compared to those outside (Liu et al. 2012). Tong et al. (2013) showed that boundary conditions and loading directions have an effect on shear band pattern in granular materials. Initiation of shear band formation can be noted in the mechanical response and exhibits different behavior under different drainage conditions. In drained conditions it is indicated by shear stress increase while in undrained condition highest rate of drop in strength and decrease in rate of dilation are the indicators (Wang and Lade 2001).

Different test methods have been used to study strain localization experimentally including plane strain and triaxial test conditions on rectangular and cylindrical samples (Oka et al. 2005, Yamamuro and Liu 2012, Yuan et al. 2013). Special methods have been used to observe shear bands formation and development (Kwak et al. 2020). Theoretical studies about strain localization in clay has been associated with strain softening induced by pore pressure generation and dissipation (Thakur 2011) creating a basis for numerical analysis of strain localization.

Strain localization in natural clay is studied numerically in this paper to investigate the effect of confining pressure and strain rate on the pattern, orientation and thickness of shear bands. An elasto-viscoplastic model proposed by Wang et al. (2015b) which was implemented into a FEM commercial software ABAQUS, is used for this research.

2 Elasto-Viscoplastic Model and FEM Simulation

2.1 Elasto-Viscoplastic Model for Natural Clay

An elasto-viscoplastic model with the ability to simulate the behaviour of natural clay proposed by Wang et al. (2015b) was used in the numerical simulations incorporating material properties of Kyuhoji marine clay (Yashima et al. 1999).

Figure 1a presents isotropic compression curve and illustrates how some of the parameters are obtained. The elastic parameters; Poisson's ratio ν and swelling index κ are inherent parameters of Cam-clay model, the former is determined based on experience and the latter from isotropic swelling test.

Critical state parameters; critical state ratio M and compression index λ are determined from triaxial compression tests and isotropic compression test respectively. Coefficient of viscosity c_0 and over-stress parameter m' are time dependent parameters determined by curve fitting in a time-dependent test (Hashiguchi 1989).

Dissipation coefficient of OCR α, dissipation coefficient of soil structure β, initial structural parameter ζ and over-consolidation ratio ω are material parameters. The first two control the rate of state variables R and R^* respectively and are determined using sub-loading concept proposed by Hashiguchi (1989). $\frac{1}{R}$ defines the OCR denoted by ω

and $\frac{1}{R^*}$ defines the structure parameter denoted by ζ, where $\omega \geq 1$ and $0 < \zeta \leq 1$. ω and ζ are determined by super-loading yield surface concept proposed by Asaoka et al. (1998). By this method, the material parameters from yield surface describing the behaviour of natural clay (structured over-consolidated state) are developed from the yield surface of Cam-clay model (normally consolidated reconstituted state) by applying similarity ratios.

θ is a calculated parameter determined by numerical integral solution. θ is time integration factor in which when $\theta = 1$ it is a fully implicit solution and when $\theta = 0.5$ it is a semi-implicit solution with the latter being at a stable state. The internal cohesion force C is determined from triaxial compression test.

Figures 1b and c present the comparison of numerical and experimental results done by Wang et al. (2015b) for undrained triaxial compression test on Kyuhoji marine clay to show the ability of the constitutive model in predicting the behaviour of natural clay. Table 1 presents the model material properties used in this study.

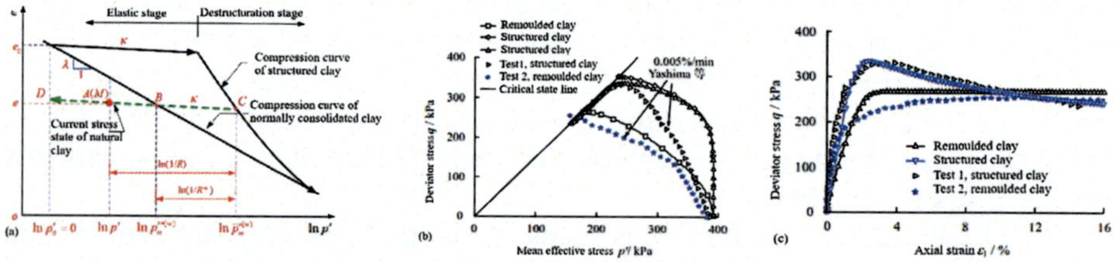

Fig. 1. (a) one-dimensional compression curve for parameter determination, undrained triaxial tests results in terms of: (b) stress path, (c) stress-strain relationship (Wang et al. 2015b).

2.2 Simulation Program

Plane strain tests with displacement control shearing were simulated to study strain localization in Kyuhoji marine under undrained conditions. Figure 2 shows the boundary and loading conditions adopted in the simulations performed to examine the effect of confining pressure and strain rate.

Confining pressures at the range of 100–500 kPa and strain rate at the range of 0.005%/min–0.5%/min for 800 eight-node plane strain quadrilateral pore pressure elements were used in the numerical simulations. In all the situations the specimen was subjected to a vertical displacement equivalent up to 40% strain during shearing stage where as the time was changed in accordance to the strain rates.

Figure 3 presents the images of the model in ABAQUS during shearing stage. When the specimen is sheared, highest strains are concentrated on the mid-section of the specimen. Shear bands initiate from the top and bottom corners of the sample and advances towards the central part until forming four crossing shear bands with two X modes.

Table 1. Model material parameters

Parameter	Poisson's ratio (ν)	Reference void ratio (e0)	Stress ratio at critical state ($M = q/p'$)	Compression index (λ)	Swelling index (κ)	Coefficient of viscosity (c_0)	Over stress parameter (m)	Dissipation coefficient of OCR (α)	Dissipation coefficient of soil structure (β)	Initial structural parameter (ζ)	Over consolidation ratio (ω)	Time integrating factor (Θ)	The internal forces of cohesion (C)
Unit						s^{-1}							kPa
Value	0.39	1.41	1.49	0.327	0.028	6.27E−11	21.5	0.5	5	1	1	0.5	10

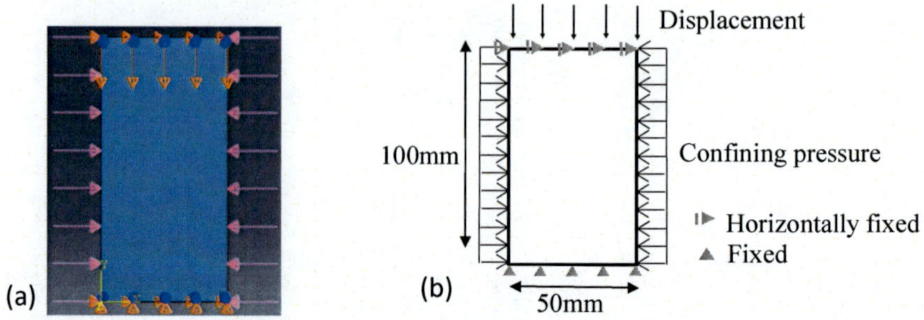

Fig. 2. Boundary and loading conditions of the model during shearing stage: (a) model as prepared in ABAQUS, (b) illustration of the boundary and loading conditions applied.

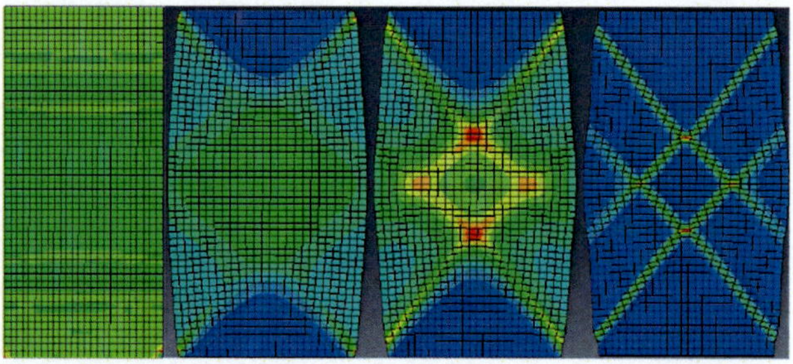

Fig. 3. Evolution of the shear band formation at different stages, left picture is at the beginning and the right one is at the end of shearing stage.

3 Finite Element Simulation of Strain Localization in Natural Clay

3.1 Effects of Confining Pressure

Stress strain responses of the overall specimen for confining pressures of 100 kPa–500 kPa are given in Fig. 4a. The general trend shows an increase in peak stress with increase in confining pressures. Very slight strain softening is shown in all cases after reaching peak stress.

Effective stress path of elements inside and outside shear bands are shown in Figs. 4b and c respectively. In the figures, **A** represents elements at the crossing of shear bands, **B** elements within the shear band away from the middle part of the model, **C** and **D** elements outside the shear band where **D** stands for those close to the shear band. Like in the overall specimen, increase in confining pressure increases the peak stress of elements both inside and outside the shear bands. Elements inside shear bands reaches a critical state after passing the peak stress whereas elements outside shear bands after peak stress they don't reach critical state instead experiences unloading.

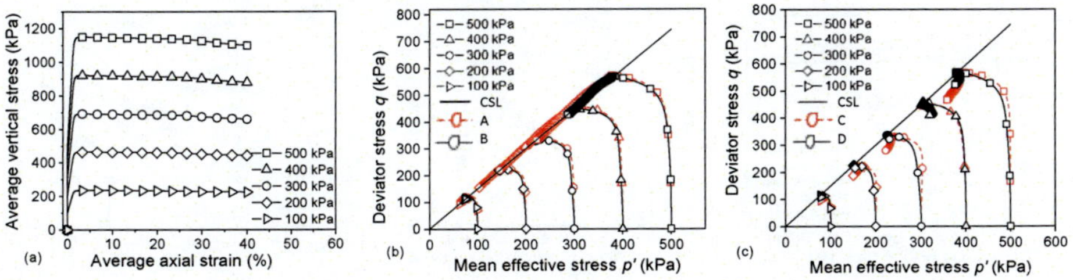

Fig. 4. Effect of confining pressure on: (a) stress strain relationship of the whole specimen, (b) stress path of shear band elements, and (c) stress path of elements outside shear bands.

Strain localization is observed for all confining pressures used in the simulation. A change in the confining pressure has no effect on the shear band pattern. Figure 5 illustrates the effect of confining pressure in shear band characteristics. It is shown that shear band inclination and thickness decrease with the increase in confining pressure. The obtained trend of results is in agreement with work by Sachan (2007) where the shear band inclination angle increased with increase in confining pressure, with the angle measured with respect to the vertical plane whereas in this paper it is measured with respect to the horizontal plane.

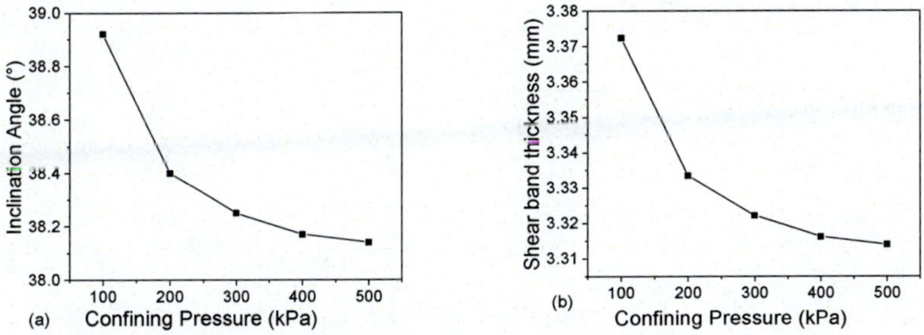

Fig. 5. Influence of confining pressure on: (a) shear band inclination angle and, (b) shear band thickness.

3.2 Effects of Strain Rate

Stress strain responses of the specimen under different strain rates are presented in Fig. 6a. The peak stress increases with an increase in strain rate. The average axial strain corresponding to peak stress is almost the same for different strain rates.

Figures 6b and c presents the effective stress path of elements inside and outside shear band respectively. An increase in strain rate increases the peak stress for elements both inside and outside the shear band. This is in agreement with explanations given by Gylland et al. (2013) that with higher strain rate the accumulation and dissipation of pore water pressure inside shear bands is higher than in slower strain rate thus experiencing higher mean effective stress and peak stress.

A change in the strain rate has no effect on the pattern of shear band, four crossing shear bands forming two X modes are observed. Figure 7 presents the effect of strain

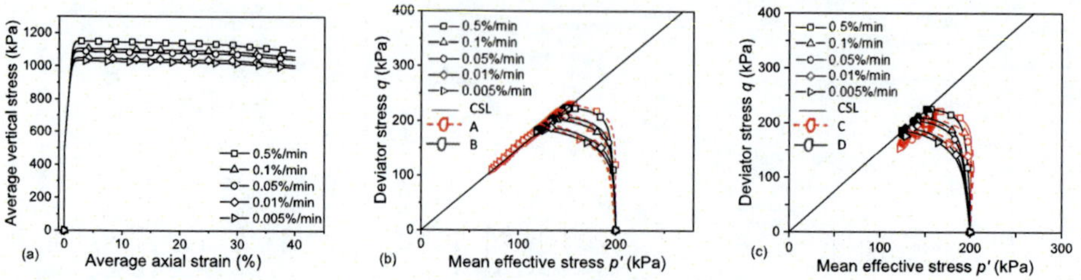

Fig. 6. Effect of strain rate on: (a) stress strain relationship of the whole specimen, (b) stress path of shear band elements and, (c) stress path of elements outside shear bands.

rate on shear band features. Figure 5a shows that, there is a slight increase in shear band inclination angle with the increase in strain rate, same trend of results was observed in experimental and numerical analysis of triaxial tests on rectangular specimens done by Oka et al. (2005). However, an opposing trend is observed in the work by Gylland et al. (2013). Figure 5b shows that with increasing strain rate shear band thickness increases opposing to results demonstrated in the work by Higo et al. (2006). In theory (Thakur 2011), slow strain rates provide sufficient time for pore water to drain out of shear band resulting to higher shear resistance and less plastic strain than with higher strain rates where a greater reduction in shear resistance and higher plastic strain is expected thus resulting to thinner shear bands.

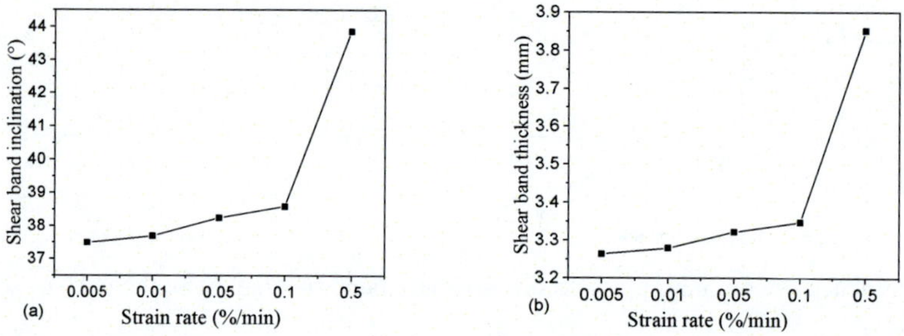

Fig. 7. Influence of strain rate on: (a) shear band inclination angle, (b) shear band thickness.

3.3 Mesh Dependency

The use of an elasto-viscoplastic model with rate dependency is one method to reduce mesh dependency (regularization) in Finite Element Analysis. To check the adequacy of the regularization method of using strain rate dependency through the use of elasto-viscoplastic model as is used in this research, the overall stress-strain relationship of the specimen with different mesh sizes was used (Oka et al. 1995).

Figure 8a shows that no mesh dependency even after reaching peak stress, a slightly higher softening rate for finer mesh is observed for axial strain greater than 25%. It can be stated with confidence that this elasto-viscoplastic model by Wang et al. (2015b) is not mesh dependent with regard to overall stress-strain response.

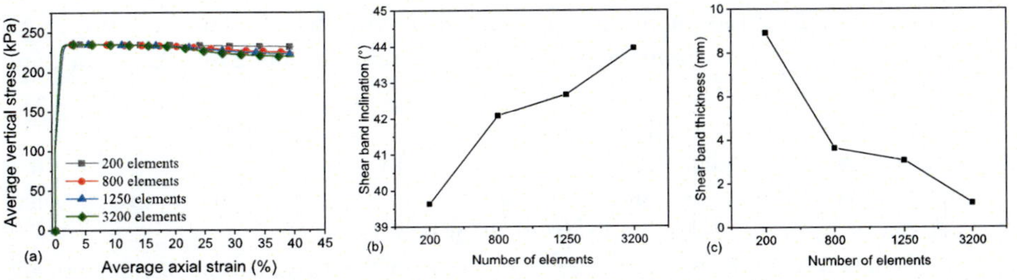

Fig. 8. Effect of mesh size on: (a) stress-strain relationship of overall specimen, (b) shear band inclination, (c) shear band thickness

In Fig. 8b and c with finer mesh, Shear band inclination angle increases while its thickness decreases. These measurements are however pointed out not to be a reliable way to check mesh dependency of a constitutive model in FEM (Oka et al. 1995). Simulations using 200 elements produce two indistinct crossing shear bands forming X pattern while 800, 1250 and 3200 elements produce four crossing shear bands that are more distinct forming two X patterns of shear band type. The shear band patterns for different number of elements are illustrated in Fig. 9.

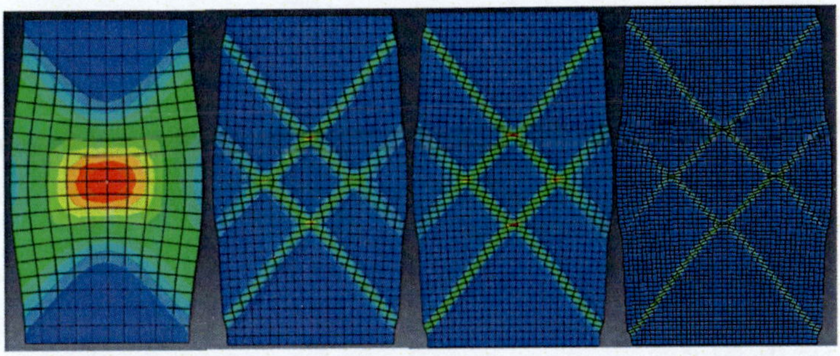

Fig. 9. Shear band patters for 200, 800, 1250, and 3200 elements (from the left)

4 Conclusions

Plane strain tests on Kyuhoji marine clay were simulated using a FEM software ABAQUS implementing an elasto-viscoplastic constitutive model to study strain localization in natural clay and the influence of confining pressure and strain rate on shear band characteristics.

The main conclusions can be drawn as follows:

i. Increase in confining pressure increases the peak stress of the overall model and individual elements. Shear band pattern is not influenced by confining pressure; however, its inclination angle and thickness decrease with the increase in confining pressure.

ii. Increase in strain rate increases the peak stress after which elements outside the shear band undergoes unloading. Increase in strain rate increases the shear band inclination angle and thickness as well.
iii. After shear band formation, elements within the shear band behave non-homogeneously. Higher peak stress is exhibited by elements located where shear bands cross. Elements outside but close to shear band show different mechanical behaviour from those that are far from shear band.
iv. Mesh dependency effect is not very apparent in the stress-strain response of the whole specimen for axial strains less than 25%. Finer mesh shows slightly higher softening rate beyond 25% axial strain. Mesh size greatly affects the shear band features in that finer mesh result in thinner shear bands with higher inclination angles.

Acknowledgments. The authors are grateful for the research funding from National Nature Science Foundation of China with Grant Nos. 51890911 and 51639008 and Postgraduate innovation research project in Tianjin University with Grant No. 2019YJSB165. The authors would also like to thank the reviewers for their comments that have greatly improved the quality of the paper.

References

Alshibli, K.A., Akbas, I.S.: Strain localization in clay: plane strain versus triaxial loading conditions. Geotech. Geo. Eng. **25**, 45–55 (2007)

Andresen, L., Jostad, H.: The role of instability and shear band localisation in triggering landslides in sensitive clays. In: Thakur, V., L'Heureux, J.-S., Locat, A. (eds.) Landslides in Sensitive Clays. ANTHR, vol. 46, pp. 179–190. Springer, Cham (2017). https://doi.org/10.1007/978-3-319-56487-6_16

Asaoka, A., Nakano, M., Noda, T.: Super loading yield surface concept for the saturated structured soils. In: Conference on Application of Numerical Methods in Geotechnical Problems, Japan, pp. 233–242 (1998)

Dao, M., Li, M.: A micromechanics study on strain-localization-induced fracture initiation in bending using crystal plasticity models. Philos. Mag. A **81**(8), 1997–2020 (2001)

Desrues, J., Besuelle, P., Lewis, H.: Strain localization in geomaterials. Geol. Soc. Lond. Spec. Publ. **289**(1), 47–73 (2007)

Gylland, A.S., Jostad, H.P., Nordal, S.: Experimental study of strain localization in sensitive clays. Acta Geotech. **9**(2), 227–240 (2013). https://doi.org/10.1007/s11440-013-0217-8

Hashiguchi, K.: Subloading surface model in unconventional plasticity. Int. J. Solids Struct. **25**(8), 917–945 (1989)

Higo, Y., Oka, F., Kodaka, T., Kimoto, S.: Three-dimensional strain localization of water-saturated and numerical simulation using an elasto-viscoplastic model. Phil. Mag. **86**(21–22), 3205–3240 (2006)

Jiang, M.J., Zhu, H.H., Li, X.M.: Strain localization analyses of idealized sands in biaxial tests by distinct element method. Front. Archit. Civ. Eng. China **4**(2), 208–222 (2010)

Jiang, M.J., Yan, H.B., Zhu, H.H., Utili, S.: Modeling shear behaviour and strain localization in cemented sands by two-dimensional distinct element method analyses. Comput. Geotech. **38**, 14–29 (2011)

Kwak, T.Y., Park, K.H., Kim, J.Y., Chung, C.K., Baek, S.H.: Shear band characterization of clayey soils with particle image velocimetry. Appl. Sci. **10**(3), 1139–1154 (2020)

Liu, X.X., Papon, A., Muhlhaus, H.: Numerical study of structural evolution in shear band. Phil. Mag. **92**(28–30), 3501–3519 (2012)

Oka, F., Adachi, T., Yashima, A.: A strain localization analysis using a viscoplastic softening model for clay. Int. J. Plast. **11**(5), 523–545 (1995)

Oka, F., Kodaka, T., Kimoto, S., Ichinose, T., Higo Y.: Strain localization of rectangular clay specimen under undrained triaxial compression conditions. In: Proceedings of the 16th International Conference on Soil Mechanics and Geotechnical Engineering, pp. 841–844. IOS Press, Amterdam (2005)

Sachan, A.: Strain localization analysis of cohesive soils using digital imaging (DIA) technique. In: 1st Indian Young Geotechnical Engineers Conference, Hyderabad, India (2007)

Tang, H.X., Dong, Y., Wang, T., Dong, Y.F.: Simulation of strain localization with discrete element-Cosserat continuum finite element two scale method for granular materials. J. Mech. Phys. Solids **122**, 450–471 (2019)

Thakur, V.: Numerically observed shear bands in soft sensitive clays. Geomech. Geoeng. Int. J. **6**, 131–146 (2011)

Tong, Z.X., Zhang, L.W., Zhou, M.: DEM simulation of biaxial compression experiments of inherently anisotropic granular materials and the boundary effects. J. Appl. Math. **2013**, 1–13 (2013)

Wang, Q., Lade, P.V.: Shear banding in true triaxial tests and its effects on failure in sand. J. Eng. Mech. **127**(8), 754–761 (2001)

Wang, Z.C., Jiang, M.J., Shen, Z.F., Chen, S.L., Cai, J.: A new method for establishing elastic-viscoplastic constitutive model of clay. In: Geomechanics from Micro to Macro, pp. 699–704 (2015a)

Wang, Z.C., Jiang, M.J., Chen, S.L., Cai, J.: An elasto-viscoplastic constitutive model and its stress aligorithm based on super-loading yield surface. Rock Soil Mech. **37**(2), 357–366 (2015b). (in Chinese)

Yamamuro, J.A., Liu, Y.: Shear Banding in Clay under Axisymmetric Triaxial Stress Conditions. Geocongress 2012, Geotechnical Special Publications, pp. 1096–1105 (2012)

Yashima, A., Shigematsu, H., Oka, F., Nagaya, J.: Mechanical behaviour and micro-structure of Osaka upper-most Pleistocene marine clay. In: Proceedings of the Japan Society of Civil Engineers, 217–229 (1999). (in Japanese)

Yuan, J.Y., Zhang, Q.H., Li, B., Zhao, X.H.: Experimental analysis of shear band formation in plane strain tests on Shanghai silty clay. Bull. Eng. Geol. Environ. **72**, 107–114 (2013)

area, as differential collapse deformations in loess often result in cracks in pavement. Therefore, the evaluation of settlements due to traffic load and partial saturation is of great significance in loess area.

In order to predict permanent settlement of soft subsoil induced by traffic load, the elastoplastic models and empirical equations have been commonly used. As the traffic loading is a large number of cycles, the complicated elastoplastic model is no longer suitable for the prediction of permanent settlement. The empirical equation between the cumulative plastic strain and its main influencing factors based on laboratory test data is usually adopted. Compared to the complicated elastoplastic model, the empirical model abandons some unnecessary characteristics, such as the hysteresis characteristics, and directly describes the long-term permanent deformation behavior of soils. The power equation or logarithmic function is always used to calculate accumulated strain increases in correspondence with the cycle number of soft soil (Monismith et al. 1975; Li and Selig 1996; Chai and Miura 2002) in which the influence of strength, deformation characteristics, the magnitude and number of applications of traffic load as well as initial static and dynamic deviatoric stress were considered. As for loess, a predictive empirical model of permanent strain as a function of cycle number, cyclic amplitude, and confining pressure was proposed by Hu et al. (2019).

The loess at natural state was often used without assessing the impact of loess hydro-collapse for the settlement calculation. Bearing in mind that the loess was intensively wetted, the calculated settlements of subgrade loess based on the general standard geomechanical expressions are generally more or less conservative. The collapses of loess have been researched by many scholars. As presented by Burland (1965), the mechanism of the collapse is the break-up of the stability between inter-particle contacts. The suction decreases which induced slippage and distortion of the grains due to wetting. There are several sophisticated numerical methods for soil collapse simulation (Sun et al. 2006).

In this paper, the settlements were calculated due to traffic loading and partial loess saturation. The calculated subgrade settlement was compared with the observed settlement, and some measures were proposed in order to ensure the safe operation of the subgrade based on the settlements.

2 Settlement Calculation Due to Traffic Loading

Reliable estimation of soil long-term settlement is very important, especially for very heavy traffic, because it affected the type of foundation and the costs. For this reason, it is significant for the calculation of soil settlement induced by traffic loading. Many studies have focused on the settlement of granular materials, soft soils (Monismith et al. 1975; Li and Selig 1996). Based on a number of cyclic triaxial tests, the development of permanent strain in loess subjected to traffic loading has been given by Hu et al. (2019). The empirical formula for accumulated strain by Hu et al. (2019) was used to predict loess settlement due to traffic loading in this paper. For the practical settlement calculation, the loess compressibility at natural state was used. The static deviatoric stress, confining stress and dynamic deviatoric stress are obtained by using the static finite element method. The permanent settlement then can be calculated by the layerwise summation method.

The analysis is National Highway 310 move southward engineering which located in Yangwa, Shaanxi Province, a typical loess area on the Chinese Loess Plateau. The National Highway 310 is a two-way road with six lanes. The total thickness of the road embankment is about 1.8 m. The input parameters for the finite element analysis are listed in Table 1 according to the Chinese General Code for Design of High Way Bridge and Culverts (JTG D60-2004). The loess subgrade and corresponding geotechnical parameters of the loess were listed in Table 2 according to USCS (ASTM Standard D2487) and the geotechnical investigation report.

Table 1. Input parameters of the road embankment

Type	Thickness (m)	Modulus of compression (MPa)	Unit weight (kN/m^3)	Poisson ratio
Asphalt concrete	0.15	1200	22	0.2
Lime-fly ash gravel	0.35	1350	20	0.2
Lime treated loess subbase	0.3	50	20	0.2
Plain filling roadbed	1.0	80	18	0.25

Table 2. Geotechnical parameters of the soil layers

Layer	USCS	Layer depth (m)	Specific gravity	Dry density (g/cm^3)	Nature moisture (%)	Liquid limit (%)	Plasticity index	Cohesion (kPa)	Internal friction angle (°)
①loess	CL	7.8	2.69	1.41	14.9	24.7	7.9	11.8	24.2
②silty loess	CL	10.0	2.68	1.41	15.4	22.8	6.9	12.3	24.8
③silt	CL	5.8	2.69	1.61	20.0	24.8	7.8	15.9	23.6
④silty loess	CL	19.2	2.70	1.67	20.7	25.2	8.0	16.4	21.9

A total traffic load of 200 kN is considered, among which the front axle load is 70 kN and rear axle load is 130 kN according to JTG D60-2004. The load carried by each axle is simplified to two-point loads that act on the road surface with a distance of 1.8 m. The front axle and rear axle load have a distance of 4.0 m. According to the measured traffic volume of freight car and without considering the effects of lane number, direction and number of axles, the approximate number of traffic load applied for the analysis was considered as 760000, 1600000, 2500000, 4500000, 6800000 and 9400000 in 1 year, 2 years, 3 years, 5 years, 7 years and 9 years, respectively. Due to the symmetry of finite element model, only half of the model is analyzed, and its 3D finite element model is shown in Fig. 1(b). As described in Fig. 1(a), the traffic load is a three-dimensional distribution pattern, which would lead to three-dimensional stress in subgrade. So a 3D finite element model was used instead of 2D plane strain model. As shown in Fig. 1(b), the soil subgrade is modeled 50 m in width, 45 m in length. The thickness of road embankment is about 1.8 m, as listed in Table 1. Compression of soils is taken into account from layer ① to layer ④ and the corresponding soil depth of compressive layers is 42.8 m in total.

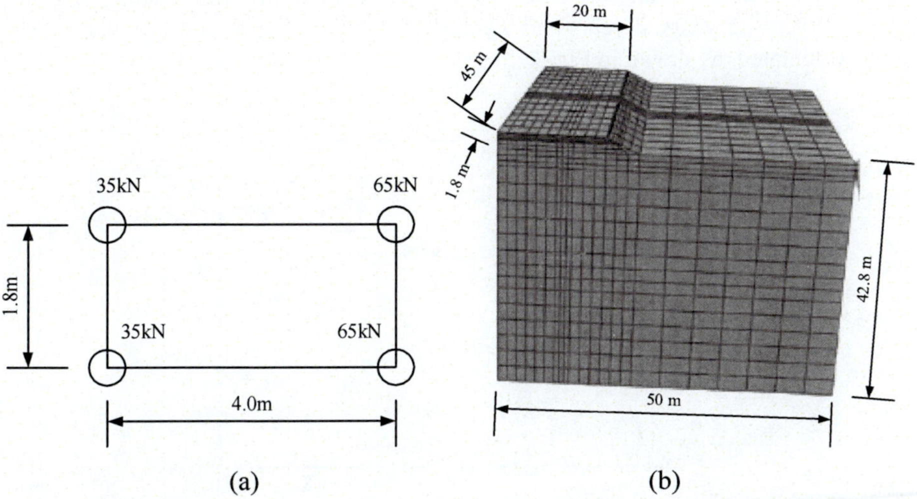

Fig. 1. The finite element model: (a) Traffic load distribution; (b) 3D finite element model

The stresses at different points along the depth under the rear axle caused by soil weight and road embankment weight were calculated through a static finite element analysis, and then the average effective consolidation pressure σ_c can be determined. The undrained dynamic deviatoric stress σ_d was determined through the stresses at different points along the depth under the rear axle caused by the traffic loads using a quasi-static finite element analysis.

According to the stress state of different soil layers below the road embankment, the cumulative deformation of each soil layer can be obtained by

$$S_t = \sum_{i=1}^{n} \varepsilon_{pi} h \tag{1}$$

where ε_{pi} is the cumulative axial strain of the ith soil layer, h_i is the thickness of the ith soil layer, and n is the number of soil layer. According to Hu et al. (2019), the empirical equation of the cumulative axial strain subjected to long-term traffic loading can be given by

$$\varepsilon_p = \left(0.03912\frac{\sigma_d}{\sigma_c} + 0.2544\frac{\sigma_d}{p_a}\right) + \left(0.0023\frac{\sigma_d^2}{\sigma_c^2} + 0.03718\frac{\sigma_d^2}{p_a^2}\right) \times \lg N \tag{2}$$

where σ_c, σ_d and p_a denote the confining pressure, dynamic deviatoric stress and atmospheric pressure ($p_a = 101.4$ kPa), respectively. The confining pressure σ_c in this paper refers to the stress caused by soil selfweight and weight of the road embankment, as described by Hu et al. (2019). The applied traffic load only generates cyclic deviatoric stress σ_d. So the confining pressure σ_c keeps unchanged with applied traffic load. For simplicity, the stress level in the middle of each soil layer is regarded as the stress level of the soil layer. Then, the cumulative axial strain of each soil layer induced by traffic loading can be obtained and the cumulative settlement due to traffic loading at 1 year, 2 years, 3 years, 5 years, 7 years and 9 years after the road open to traffic respectively can be calculated, as shown in Fig. 2.

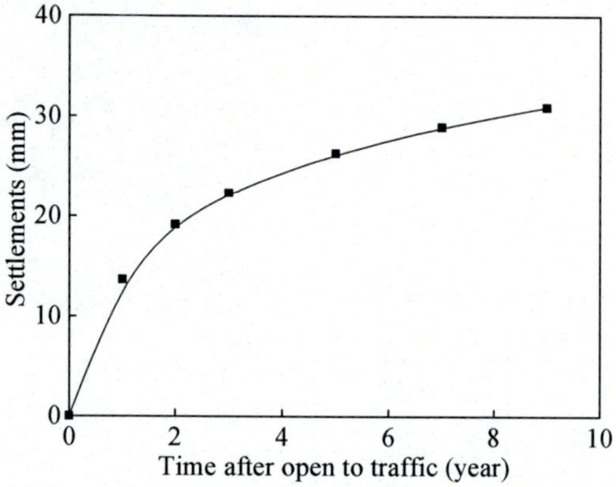

Fig. 2. Cumulative settlements due to traffic loading after the road open to traffic

3 Settlement Calculation Due to Partial Loess Saturation

For settlement monitoring of the subgrade, 6 displacement transducers were installed below the embankment. The displacement transducers were installed in a typical road vertical section. The measurement points were embedded under the surface of the embankment and located in the center of each lane, as shown in Fig. 3(a). Figure 3(b) shows the average long-term permanent settlement observed at 6 measured points. As indicated in Fig. 3(b), based on the permanent settlement at 9 years after the road open, the average value of the last observed permanent settlement was 91.8 mm, and it was 60.9 mm (1.98 times) higher than the calculated value. As a consequence, the question arose regarding the cause of higher settlements and the settlement rate.

Since the calculated settlements based on empirical equation and 3D finite element calculation are generally more or less conservative, noticing that the loess was intensively wetted, it was assumed that the major reason of higher settlements is the loess partial hydro-collapse.

It is well known that the soil moisture has a significant influence on the loess partial hydro-collapse. As shown in Table 2, the moisture of the loess grows with depth. On the other hand, long and intense rainfall can significantly increase the soil moisture in

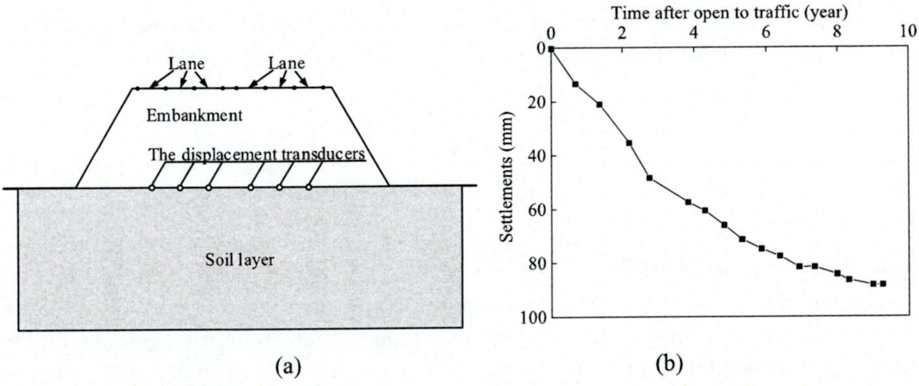

Fig. 3. Long-term permanent settlements: (a) the locations of six displacement transducers; (b) measured average long-term permanent settlements

the unprotected subgrade, which will also lead to strength reduction and compressibility and settlements increment. The parameters for the settlement calculation due to partial loess saturation were based on the double oedometer test. The samples were taken from each point every 4.0 m under the road embankment and a test result from 4.0 m below the road embankment was shown in Fig. 4. The loess collapse coefficient at each vertical pressure is obtained by Eq. (3).

$$\delta_s = \varepsilon_h = \frac{e_{\mathrm{dry}} - e_{\mathrm{sat}}}{1 + e_0} \tag{3}$$

where ε_h is the vertical deformation due to hydro-collapse, e_{dry}, e_{sat} and e_0 are void ratio before and after saturation and initial void ratio.

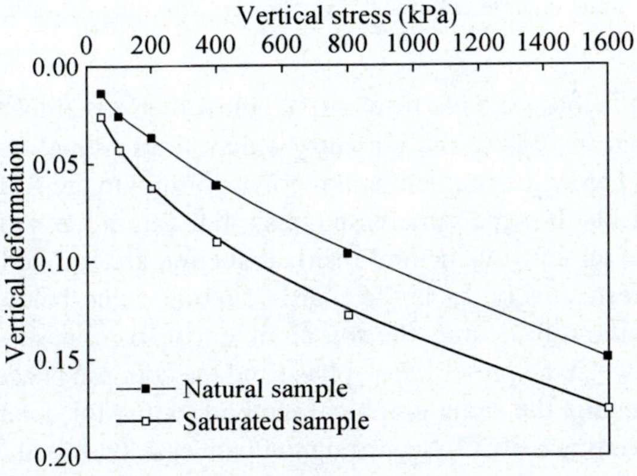

Fig. 4. Double oedometer test on loess sample from 4.0 m below the road embankment

According to the discussion given by Pereira and Fredlund (2000), the collapse soil would remain at constant volume in the pre-collapse phase and stop collapsing in the post-collapse phase, as shown in Fig. 6 (Pereira and Fredlund 2000). Fredlund (2000) gave that the void ratio at any value of matric suction can be calculated as

$$\begin{array}{ll} e = e_0 & s > s_0 \\ e = e_0 - \lambda_s \ln\left(\frac{s_0}{s}\right) & s_f \le s \le s_0 \\ e = e_f & 0 \le s \le s_f \end{array} \tag{4}$$

where e, e_0 and e_f are the void ratio, initial void ratio and final void ratio after collapse respectively, s, s_0 and s_f denote the matric suction, the critical suction and final suction respectively. Fredlund (2000) introduced three parameters to characterize the collapse behavior: the critical suction value, s_0, below which the soil collapses significantly; the stiffness parameter for changes in matric suction, λ_s, which is assumed as constant in the e-ln$s$ plane; and the final suction, s_f, as shown in Fig. 5 (Pereira and Fredlund 2000). The Soil-Water Characteristic Curve (SWCC), s_0 and λ_s are required to predict the void ratio at any value of matric suction provided that the initial void ratio, e_0 is known.

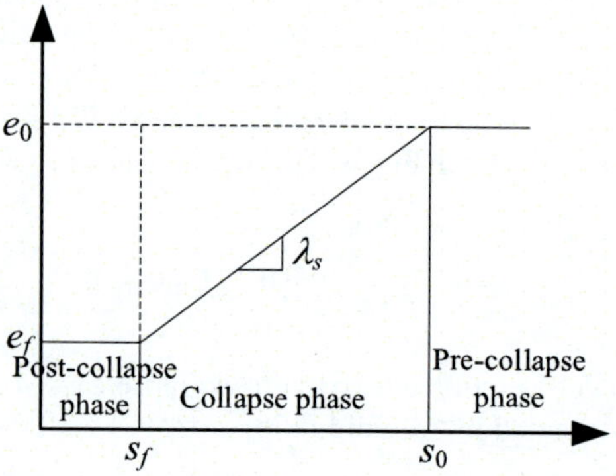

Fig. 5. Volume change behavior in various collapse phases during wetting

The SWCC of the loess sample from 4.0 m below the road embankment was shown in Fig. 6. As shown in Fig. 6, the air entry value and residual suction of the loess sample from 4.0 m below the road embankment according to the SWCC is 3.7 kPa and 111.9 kPa, respectively. It is the same with the similar nature loess in the Loess Plateau of China, where the air entry value and residual suction are variated from 2.0–8.5 kPa and 90–320 kPa, respectively. s_0 is the matric suction value below which there is a sudden decrease in the soil volume, the soil commences to collapse apparently. It is the boundary value between the pre-collapse phase and the collapse phase. It is suggested to be determined following the same graphical method to that for determining the water-entry value on the wetting SWCC. According to Pham et al. (2003), the water-entry value on the wetting curve, ψ_w, is the critical matric suction at which the soil water content starts to increase significantly during the wetting process. s_f is assumed to correspond to 85% degree of saturation on the SWCC not only because full collapse is observed to complete at approximately 85% degree of saturation, but also because the capillary saturation zone is found to be in the range from 80–90% to 100% degree of saturation.

This assumption simplifies the model, while it requires the wetting SWCC for estimating s_f. s_f can be regarded as the air entry value of SWCC after collapse. As the post-phase of loess often occurs within a small range of suction (El-Ehwany and Houston 1989), the s_f is assumed to be 1 kPa for simplicity. λ_s is the slope on the e-ln$s$ plane in the collapse phase. It is assumed to be constant, which permits the proposed equation being presented in integrated form instead of incremental form while may not accurately represent the decrease of void ratio with respect to logarithmic matric suction in the collapse phase. According to the definition, after s_0 and s_f is determined, the collapse rate λ_s can be obtained by the collapse coefficient δ_s as follows

$$\lambda_s = \frac{e_0 - e_f}{\ln s_0 - \ln s_f} = \frac{\delta_s(1 + e_0)}{\ln s_0} \tag{5}$$

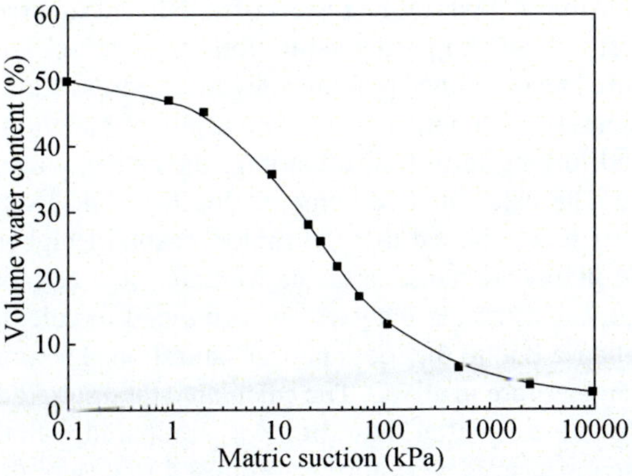

Fig. 6. The wetting soil-water characteristic curves (SWCC)

After these parameters were determined, the void ratio during wetting at any value of matric suction can be obtained through Eq. (4) and the corresponding settlements due to partial saturation can be calculated by

$$s_p = \frac{e_0 - e}{1 + e_0} H_0 \tag{6}$$

where H_0 is the height of the soil layer. The calculated settlements due to partial loess saturation by the model are shown in Table 3. As shown in Table 3, the total settlement of the subgrade caused by hydro-collapse due to the partial loess saturation is 44.5 mm.

The total settlements of the subgrade include settlements due to traffic loading and partial collapse and the calculated value is 75.4 mm, which is also 15.4 mm smaller than the observed value. In order to prevent further problems, the following measures were proposed: (1) Continue to measure settlements once a month. Afterwards, the measured frequency could be decreased depending on the settlements. (2) Change the current design of the embankment draining of the water, with a surface discharge into gutterway.

Table 3. The calculated settlement due to partial loess saturation

Depth (m)	δ_s	s_0(kPa)	e_0	λ_s	s	e	s_p
4	0.029	72	1.096	0.0142	34	1.085	20.4
8	0.021	72	1.06	0.0101	39	1.054	12.0
12	0.014	60	1.022	0.0069	39	1.019	5.9
16	0.012	60	0.989	0.0058	40	0.987	4.8
17.8	0.009	60	0.98	0.0044	42	0.978	1.4

4 Conclusions

This paper established the analysis of the results of calculated subgrade settlement in the loess area, due to traffic load and partial saturation of the loess. The average effective consolidation pressure and undrained dynamic deviatoric stress were calculated through 3D finite element analysis. Combined with the empirical equation of the cumulative axial strain subjected to long-term traffic loading, the settlements due to traffic loading were calculated. The calculated settlements due to traffic loading were less than observed value, and it was assumed that the major reason of higher settlements is the loess partial hydro-collapse. A model aims at predicting the collapse behavior during wetting based on the experimental data was incorporated to calculate the settlements caused by hydro-collapse due to the loess partial saturation. The model is quite simple since only three parameters are involved. The calculated total subgrade settlement due to both traffic loading and loess partial saturation was compared with the observed values for verification. Some measures were further proposed to ensure the safe operation of the subgrade based on the settlements.

References

ASTM-D2487: Standard Practice for Classification of Soils for Engineering Purposes (Unified Soil Classification System). West Conshohocken (2017)

Burland, J.B.: Some aspects of the mechanical behavior of partly saturated soils. In: Moisture Equilibria and Moisture Changes in Soils Beneath Covered Areas. Butterworths, Sydney, Australia, pp. 270–278 (1965)

Chai, J.C., Miura, N.: Traffic-load-induced permanent deformation of road on soft subsoil. J. Geotech. Geoenviron. Eng. **128**(11), 907–916 (2002)

El-Ehwany, M., Houston, S.L.: Settlement and moisture movement in collapsible soils. J. Geotech. Eng. **116**(10), 1521–1535 (1989)

Fredlund, M.D.: The role of unsaturated soil property functions in the practice of unsaturated soil mechanics. University of Saskatchewan (2000)

Hu, Z.P., Wang, R., Ren, X., Wei, X.N., Wang, Q.: Permanent deformation behavior of compacted loess under long-term traffic loading. J. Mater. Civ. Eng. **31**(8), 04019157 (2019)

Li, D., Selig, E.T.: Cumulative plastic deformation for fine-grained subgrade soils. J. Geotech. Eng. **122**(12), 1006–1013 (1996)

Li, P.: Prediction of the wetting-induced collapse behaviour of unsaturated soils. Chang'an University, Xi'an (2016)

Monismith, C.L., Ogawa, N., Freeme, C.R.: Permanent deformation characteristics of subgrade soils due to repeated loading. Transport. Res. Rec. J. Transport. Res. Board **537**, 1–17 (1975)

Pereira, F., Fredlund, D.G.: Volume change behavior of collapsible compacted gneiss soil. J. Geotech. Geoenviron. Eng. **126**(10), 907–916 (2000)

Pham, H.Q., Fredlund, D.G., Barbour, S.L.: A practical hysteresis model for the soil-water characteristic curve for soils with negligible volume change. Géotechnique **53**(2), 293–298 (2003)

Sun, D.A., Sheng, D., Sloan, S.W.: Elastoplastic modelling of hydraulic and stress-strain behavior of unsaturated soil. Mech. Mater. **39**, 212–221 (2006)

way of excavation, backfill, and transformation has caused great changes to the topography. These destroy the fragile loess environment. It is becoming widely accepted that the benefits of fiber-reinforced soil to control the slope stability by increasing the shear resistance of soil. Lian et al. (2019) investigated the shear strength of loess reinforced with root by performing consolidated-undrained triaxial tests and found that the increase of internal cohesion contributes the most to the increase in strength of reinforced specimen. Dinh et al. (2020) investigated the characteristics of carbon fiber textile-reinforced mortar (TRM) under direct shear considering the main testing parameters including the fiber reinforcement ratio of carbon fiber textile, the surface treatment used between mortar matrix and the fiber textile, and the inclination angle of the fiber filaments. The test results showed that the reinforcement affect the failure modes and the shear performance of tested specimens in terms of cracking and ultimate shear stress, crack deformation, and residual shear stress in the pre- and post-peak stages. Peerun et al. (2019) studied the mechanical behaviour of sand and rock spoils of metagreywacke and shale origins using a creative combination of tools including a transparent shear box and particle image velocimetry (PIV) technique. This interpretation measure effectively demonstrated how particle shapes and angularity, and the effect of particle dilation and compression under shear can affect the strength parameters. Tan et al. (2019) carried out direct shear tests on fallow soil specimens cultivated using three-dimensional geomats and Bermuda grass, namely, root-geomat reinforced soil (RMS), and the test results showed that the shear strength of RMS could be considerably improved by the combined reinforcement with grass root and geomats. The improvement seemed to have a more pronounced effect at a low vertical stress level, whereas the soil cohesion was increased considerably, but the influence on the friction angle was negligible. Fatahi et al. (2012) and Nguyen and Fatahi (2016) studied the mechanical characteristics of soft clay treated with fiber and cement and found that the fiber contributes to the increase in peak strength. The addition of fiber increases the residual strength and changes the brittle behavior of cement treated with clay to a more ductile material. The tensile strength of cement treated clay is increased with increase of carpet and steel fiber.

In addition to laboratory tests, studies on the shearing behaviour of geomaterials were also performed using numerical and field observation approaches. Wang et al. (2019b) studied the effect of the grain size distribution of a sandy soil on the shearing behaviour at the soil-structure interface using a modified direct shear apparatus and the PFC2D software. The shearing resistance provided by the force chain of the smaller particles (0.21 mm diameter) showed lower values than that formed by larger particles (0.79 mm diameter). Tabarsa et al. (2018) investigated the potential for effective loess stabilisation using nanoclay, namely, an engineered nanomaterial, both in the laboratory and in the field at the Gonbad dam irrigation channel site. Varying fractions of nanoclay ranging from 0.2% to 3% by mass were added to the loess soil. Results indicated that loess soils that had been stabilised with 2% nanoclay showed remarkable improvement and were in line with the field observation. The above studies primarily explore the potential for insertion of plant root and barley straw to improve the shear strength of natural soils either using direct shear test or triaxial shear test towards controlling strata erosion. However, studies exploring the shearing behaviour versus time for reinforced loess soils when subjected to different magnitudes of loads and detailing the behind mechanism of strength improvement is remarkably limited.

Nowadays a significant body of researches has sought an environment-friendly construction material towards achieving sustainable development (Zhang et al. 2010; Sivakumar Babu and Raja Jaladurgam 2014; Suksiripattanapong et al. 2017; Yoobanpot et al. 2017; Kua et al. 2017; Xu et al. 2019). Loess and wheat straw which is deemed as one of the agricultural solid wastes obtained after harvesting of wheat grains are easily accessible in the Chinese Loess Plateau. In contrast to a high utilisation efficiency of such agricultural solid waste of 90.7% in Japan, relatively low utilisation efficiency of 70.6% in China has been caused pronounced impacts to surrounding environment. The agricultural solid waste produced annually can reach more than 4 billion tons in China and the waste straw occupies 17.5% of the total amount. This fact exemplifies the importance of reducing, recycling, and reusing the waste straw in relation to sustainable development and protection of the environment. Despite many countermeasures available (Hungr 1995, 2006; Modoni et al. 2006; Mangeney et al. 2010; Crosta et al. 2008, 2013; Modoni and Bzòwka 2012; Xu et al. 2012, 2013; Shen et al. Shen et al, 2013a, b, c; Modoni et al. 2016; Shen et al. 2017; Cheng et al. 2018a, b, c; Duan et al. 2018; Cheng et al. 2019; Gu et al. 2019), loess-waste straw mixture may be considered as an alternative backfill material not only to reduce agricultural solid waste, but to prevent instability of loess body in hilly-gullied regions of northwest China towards protecting environment and reducing the potential of landslide hazard (Zhuang et al. 2017). Literature to discuss the interactive behaviour between soil and fiber and the shear strength enhancement mechanism however is remarkably limited.

The objectives of this paper are: (a) to document the baseline strength parameters of Lantian loess and Lantian loess reinforced with the agricultural waste straw under displacement-controlled direct shear for three normal pressures, (b) to investigate the temporal relation of horizontal and vertical displacement of the same soil and reinforced one under stress-controlled direct shear for a given normal pressure, and (c) to explore the potential for the use of loess-waste straw mixture as an alternative backfill material to functionalise the agricultural waste straw and improve the stability of loess against shallow slope failure.

2 Materials and Methods

2.1 Testing Materials

Natural loess block sample of 1 m^3 from Lantian (Fig. 1), about 22 km southeast of Xi'an, Shaanxi Province of China, was utilised for preparing a series of direct shear specimens. Details in relation to their preparation are given as follows:

(1) Dry the Lantian loess by the temperature of 105 °C for 24 h; measure the natural water content.
(2) Crush the dried loess; sieve the dried loess with a sieve with 2 mm opening ensuring all the fines to be included despite the presence of a small amount of sand.
(3) Pave the dried loess in a disk; trim waste straws to 40 mm long and treat them with boiled water for the convenience sake of specimen preparation (Fig. 2).

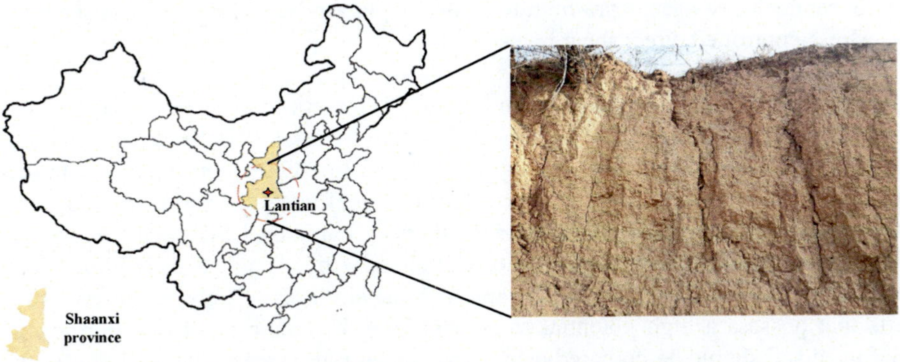

Fig. 1. Sampling location (Lantian, about 22 km southeast of Xi'an)

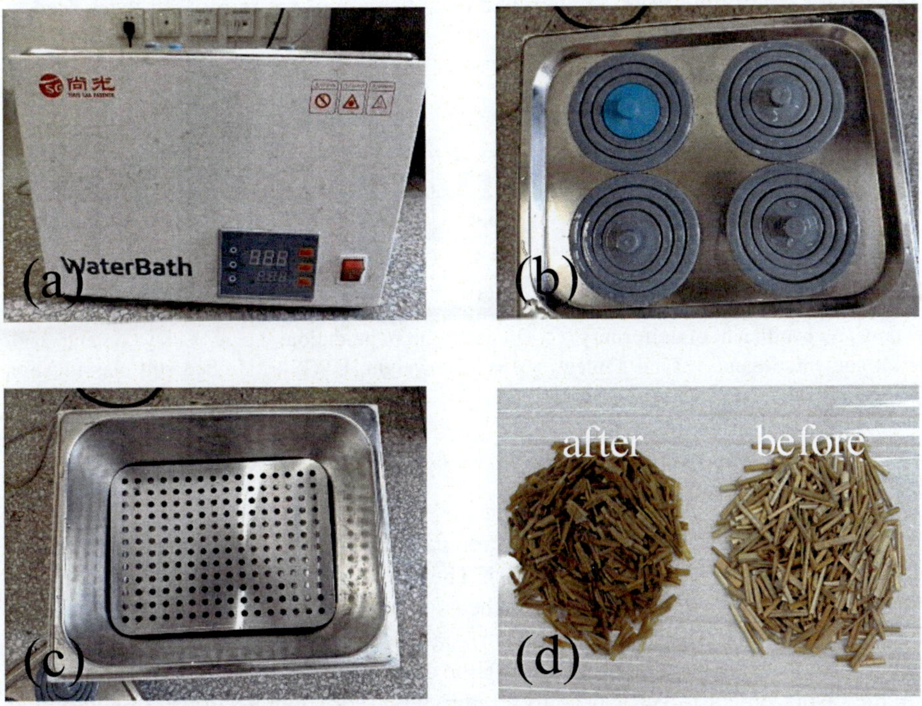

Fig. 2. Water bath apparatus: (a) control panel, (b) sub water tanks, (c) heater, and (d) waste straw before and after the boiled water treatment

(4) Calculate the required weight of loess and amount of water for a single specimen in accordance with the unit weight γ_d of 14 kN/m^3.
(5) Place the loess-waste straw mixture in a moist chamber for two days achieving the water content uniform in soil.
(6) Compact the loess-waste straw mixture in a square shear box by three layers.

(7) Shear the loess-waste straw mixture under displacement-controlled direct shear and stress-controlled direct shear respectively.

The physical properties for Lantian loess are summarised in Table 1. The particle-size distribution curves are shown in Fig. 3. The Lantian loess classifies as low plasticity silt (ML) with the silt content varying from 81.1% to 84%, according to the unified soil classification system (USCS). The uniformity coefficient C_u varies in 9.26–15.26 although the majority of the natural loess shares a common USCS symbol. Further, it is not likely to ensure the uniformity of the straw fiber applied to reinforce loess soil in real practice. The possible means to achieve this is to only reinforce critical parts that possess a high potential to subject large hydraulic gradients. Further, the reinforced soil should be prepared using a manufacturing process that is similar to that of geogrid, widening its application horizon in future practice. Moreover, a preliminary compaction test can provide the optimal water content and maximum dry density which are used for controlling water content and densification of the reinforced loess used in land reclamation.

Table 1. Physical properties of Lantian loess

Sample ID	C_u	C_c	%clay	%silt	%sand	γ_d (kN/m^3)	USCS symbol
S1a	9.26	1.94	10.0	83.4	6.6	13.72	ML
S1b	10.48	2.13	11.0	81.1	7.9	13.60	ML
S1c	15.26	2.79	10.0	84.0	6.0	13.83	ML

Note: C_u = coefficient of uniformity; C_c = coefficient of gradation; %clay = clay content; %silt = silt content; %sand = sand content; γ_d = unit weight; USCS = Unified soil classification system.

2.2 Testing Apparatus and Methods

The shear box consists of upper and lower parts and they have equal dimensions of 305 mm × 305 mm × 100 mm (Fig. 4). The shear force imposed to a tested specimen can reach up to 44.5 kN. The direct shear apparatus is equipped with micro-stepper motors for controlling both vertical and horizontal load and displacement. Two displacement transducers are capable of measuring the displacement in horizontal up to 180 mm and the displacement in vertical up to 90 mm, with 0.002 mm accuracy. In addition to displacement-controlled direct shear, the direct shear apparatus is also capable of imposing stress-controlled direct shear to a tested specimen, with the shearing rate varying from 0.00003 mm/min to 7.5 mm/min.

To document the baseline shear strength parameters, the specimens when subjected to displacement-controlled direct shear were prepared at the water contents ω of 14%, 18%, and 22% respectively, with insertion mass percentage of the waste straw (termed %waste straw hereafter) corresponding to 0.3%, 0.45%, 0.6%, 0.75%, and 1.0%. The specimens were sheared at a rate of 0.8 mm/min under three normal pressures σ_n of 100 kPa, 200 kPa, and 300 kPa respectively until a maximum horizontal displacement

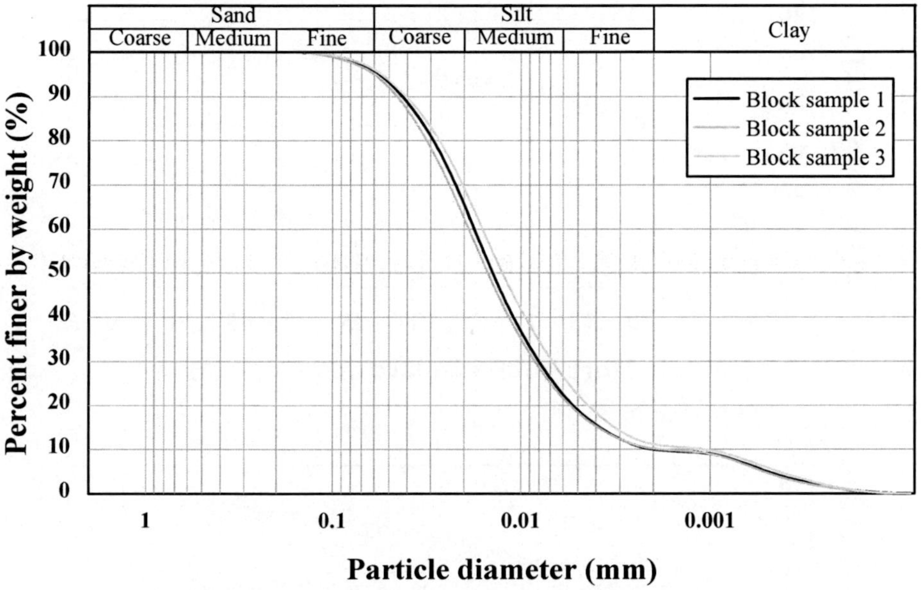

Fig. 3. Particle-size distribution curves for the Lantian loess

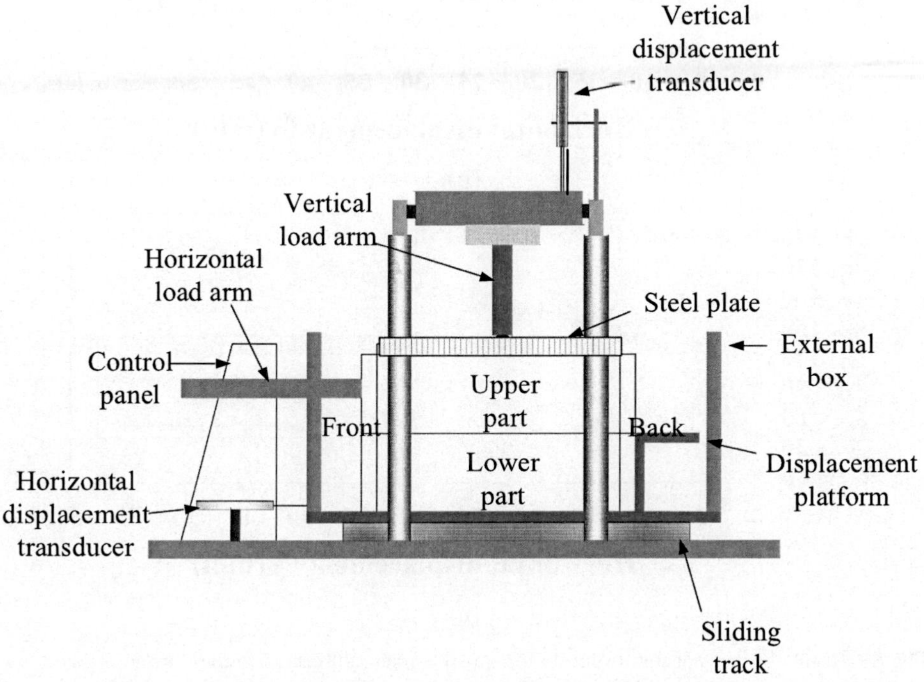

Fig. 4. Illustration of the direct shear test apparatus

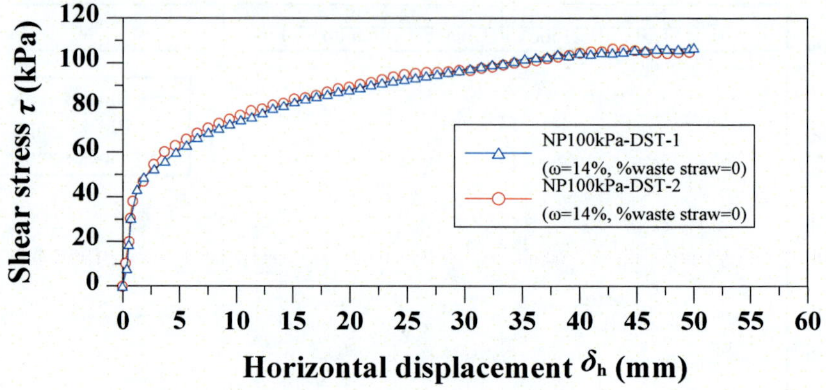

(a)

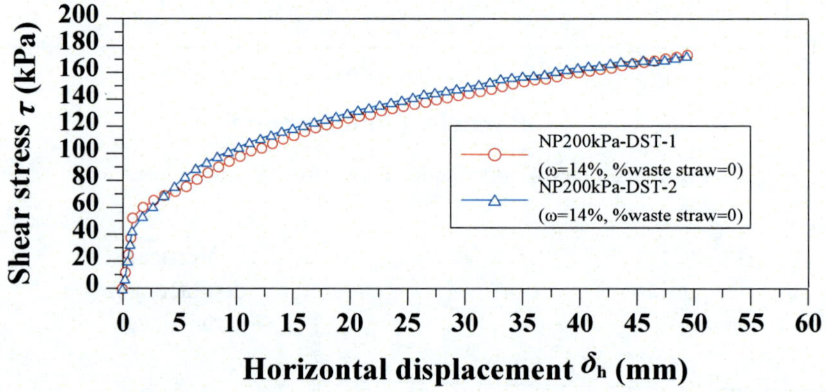

(b)

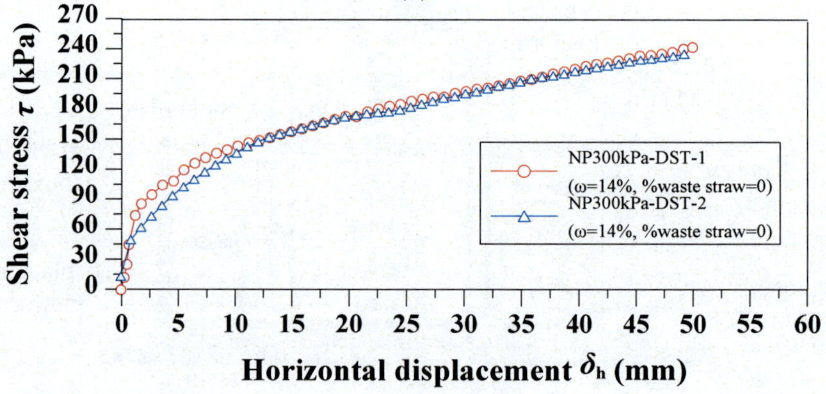

(c)

Fig. 5. Results of the repeatability tests for: (a) specimens subjected to the normal pressure σ_n of 100 kPa, (b) specimens subjected to the normal pressure σ_n of 200 kPa and (c) specimens subjected to the normal pressure σ_n of 300 kPa

$\delta_{h,max}$ of 50 mm attained. Comparison with the specimens prepared at %waste straw = 0 was performed to highlight the effect of waste straw inclusions.

Optimal testing parameters (%waste straw and water content ω) obtained from the displacement-controlled direct shear test results were then fed back to the preparation of specimens subjected to stress-controlled direct shear. The specimens were sheared under $\sigma_n = 95$ kPa until the failure of specimens attained during stress-controlled direct shear tests. The failure for a specimen was defined as a change in curvature of the relation of horizontal displacement δ_h versus elapsed time t from positive to negative and a rapid increase in δ_h value.

2.3 Testing Method Validation

Repeatability tests to check and validate the accuracy of the testing methods appeared necessary prior to the direct shear tests. Repeatability of the direct shear method was verified using six repeatability tests undertaken at the specimens prepared at %waste straw = 0 and $\omega = 14\%$ under $\sigma_n = 100$ kPa, 200 kPa, and 300 kPa respectively. The ultimate shear stress τ_{ult} was defined as the point on the curve where the curvature changes. Here an uncertainty parameter, defined as the ratio between the variation range of τ_{ult} and the average of τ_{ult}, was introduced to quantify the repeatability of these tests. The results of the repeatability tests are shown in Fig. 5 as the relation of shear stress τ versus horizontal displacement δ_h. While the failure envelopes are shown in Fig. 6. Relations of τ versus δ_h for a given σ_n exemplify the minimal difference in shear behaviour

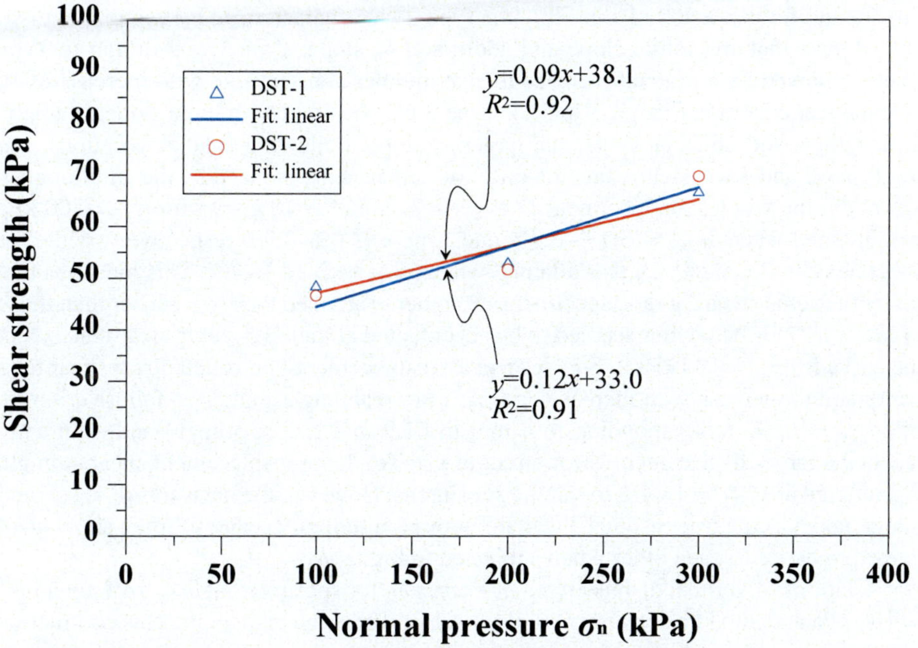

Fig. 6. Failure envelopes from the repeatability tests

between specimens, and the two failure envelopes showed good correspondence to each other. The friction angle ϕ' varied from 5.1° to 6.8°, whereas the cohesion c' varied from 38.1 kPa to 33 kPa. The uncertainty parameter in relation to the ultimate shear stress τ_{ult} was determined as up to 2.1%. The above results, therefore, verified and ensured repeatability of the testing apparatus and direct shear method.

3 Results

3.1 Shear Behaviour in Displacement-Controlled Direct Shear Tests

To document the baseline shear strength parameters, displacement-controlled direct shear tests on Lantian loess specimens at $\omega = 14\%$, 18%, and 22% respectively, with %waste straw = 0, 0.3, 0.45, 0.6, 0.75, and 1.0, were conducted under $\sigma_{\mathrm{n}} = 100$ kPa, 200 kPa, and 300 kPa respectively. For the sake of brevity, a part of the displacement-controlled direct shear test results is shown in Figs. 7, 8 and 9. It can be seen that τ increased with increase of $\delta_{\mathrm{h,rev}}$ which is defined as the ratio between the δ_{h} and 305 mm (specimen size). The increase in τ was found to be steeper for higher σ_{n}. It is noteworthy that the relations of τ versus $\delta_{\mathrm{h,rev}}$ behaved in a strain-hardening manner and were classed into three types. In the case τ increasing to an ultimate value τ_{ult} and then remaining essentially constant with increase of δ_{h} was classed as Type 1 (see the curve for $\sigma_{\mathrm{n}} = 100$ kPa in Fig. 7c). τ_{ult} was, therefore, defined by the horizontal tangent. Type 2 was described as τ increasing to a curve turning point where some factor begins to have some influence on shearing behaviour and, thereafter, gradually increasing at a constant rate with increasing δ_{h} (see the curves for $\sigma_{\mathrm{n}} = 200$ kPa and 300 kPa in Fig. 7a and for $\sigma_{\mathrm{n}} = 300$ kPa in Fig. 8a). Type 2 was characterised by two distinguishable slopes; that are, initial slope and additional δ_{h} slope. Type 3 was similar to Type 2, but followed by a gradual increase in τ at another constant rate with increase of δ_{h} (see the rest curves in Figs. 7, 8 and 9). Type 3 was featured with three slopes, namely, initial slope, additional δ_{h} slope, and larger δ_{h} slope. Following the classifications, τ_{ult} for Types 2 and 3 was defined by the onset of tangent line, induced by the additional δ_{h} slope. On the whole, for specimens at $\omega = 14\%$ and $\omega = 18\%$, τ_{ult} for $\sigma_{\mathrm{n}} = 300$ kPa was attained when $\delta_{\mathrm{h,rev}} = 3.1\%$–5.4% and $\delta_{\mathrm{h,rev}} = 3.0\%$–4.9% respectively, while for specimens at $\omega = 22\%$ τ_{ult} was attained when $\delta_{\mathrm{h,rev}} = 2.7\%$–4.7%. This indicates that under the same shearing rate, specimens of higher ω attained their τ_{ult} earlier than those of lower ω. This behaviour seemed to be accentuated at higher σ_{n}. It is clear that except the curve for $\sigma_{\mathrm{n}} = 100$ kPa in Fig. 7c, τ in all displacement-controlled direct shear tests continued to increase with increase of $\delta_{\mathrm{h,rev}}$ after reaching a suggested failure criterion of $\delta_{\mathrm{h,rev}} = 6.5\%$ (corresponding to 4 mm in 61.9 mm size specimen) and eventually reached a large-displacement shear strength τ_{ld}. The large-displacement shear strength τ_{ld} adopted at $\delta_{\mathrm{h,rev}} = 16.4\%$ for all the specimens. However, the measures of τ_{ld} should not be taken as the true residual shear strength τ_{r}, although τ_{ld} and τ_{r} often show good correspondence to each other when subjected to higher δ_{h}.

Relations of vertical displacement δ_{v} versus δ_{h} for specimens at $\omega = 14\%$ under $\sigma_{\mathrm{n}} = 100$ kPa and 300 kPa are shown in Fig. 10. Positive value of δ_{v} represents contraction of specimen, whereas negative value of δ_{v} represents expansion. In all displacement-controlled direct shear tests, δ_{v} was measured to be positive, indicating that specimen

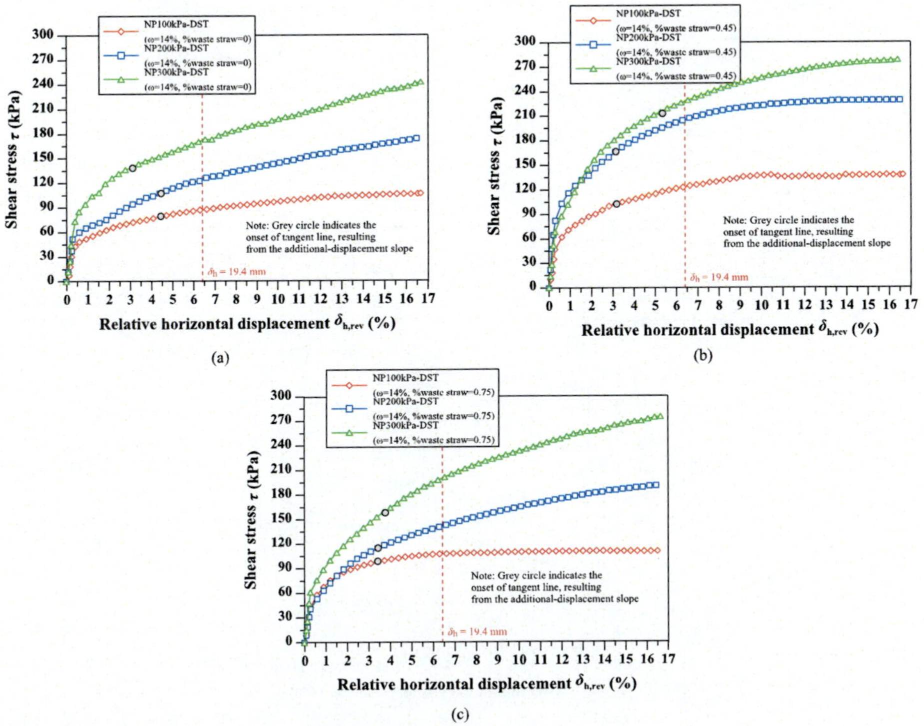

Fig. 7. Shear stress τ versus relative horizontal displacement $\delta_{h,rev}$ relation for specimens at $\omega = 14\%$ with (a) %waste straw = 0, (b) %waste straw = 0.45, and (c) %waste straw = 0.75 under the normal pressures of 100 kPa, 200 kPa, and 300 kPa

thickness decreased with increasing δ_h. It can also be observed from Fig. 10 that the magnitude of δ_v varied significantly between specimens under $\sigma_n = 100$ kPa and 300 kPa and on the whole, it was larger for specimens under $\sigma_n = 300$ kPa than under $\sigma_n = 100$ kPa. This contractive behaviour was mainly attributed to higher σ_n. Higher σ_n could also result in omittance of top platen rotation during shearing. Ghazizadeh and Bareither (2018) indicated that top platen rotation has not been discussed because this rotation issue is suppressed with higher $\sigma_n \geq 100$ kPa. However, for specimens with %waste straw = 0, 0.3, 0.75, and 1 under $\sigma_n = 100$ kPa, δ_v decreased with increase of δ_h after reaching a peak of δ_v, implying that the specimens would have behaved dilative behaviour if δ_h was large enough to develop negative value of δ_v.

3.2 Baseline Shear Strength and Optimal Percentage of Waste Straw Inclusions

Strength envelopes, based on τ_{ult} determined using the onset of tangent line resulting from the additional δ_h slope, for specimens at $\omega = 14\%$, 18%, and 22% with different %waste straw are shown in Fig. 11. Baseline shear strength parameters including c and ϕ are tabulated in Table 2, in accordance with the strength envelopes. Although cohesion c_{ld} and friction angle ϕ_{ld}, associated with large-displacement shear strength τ_{ld}, are

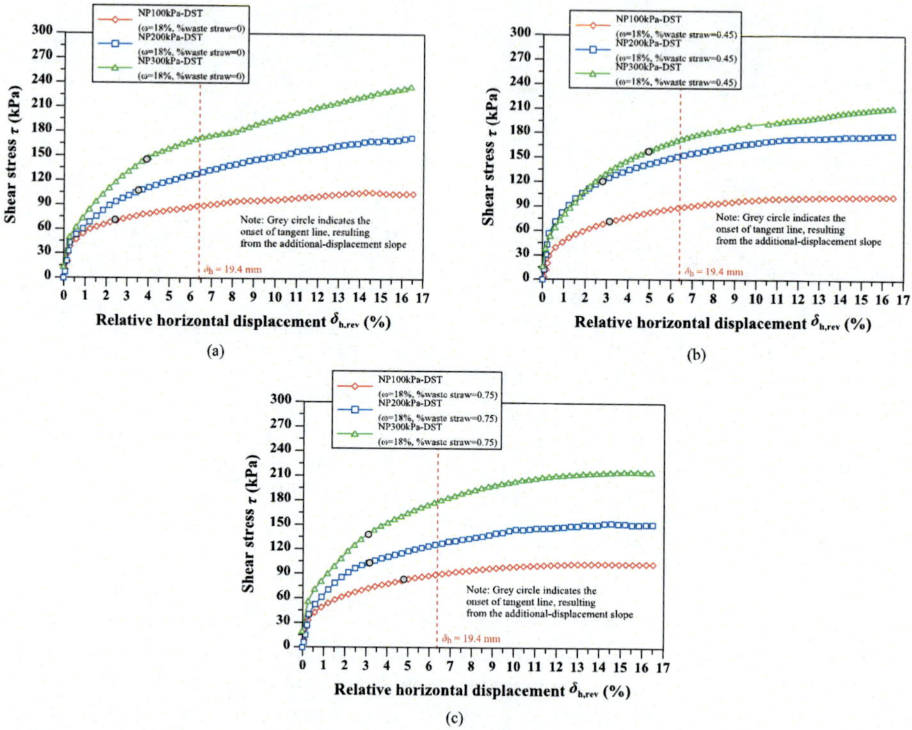

Fig. 8. Shear stress τ versus relative horizontal displacement $\delta_{h,rev}$ relation for specimens at ω = 18% with (a) %waste straw = 0, (b) %waste straw = 0.45, and (c) %waste straw = 0.75 under the normal pressures of 100 kPa, 200 kPa, and 300 kPa

tabulated in Table 3, they were not taken as the baseline shear strength parameters. It is clear that ϕ varied between specimens and was higher for specimens with higher %waste straw. ϕ appeared to ascend for a %waste straw range of 0.3–0.6 and descend for a %waste straw range of 0.6–0.75 as well as ascend again for a %waste straw range of 0.75–1.0. The γ_d values were tested close to the maximum of 15.6 kN/m^3 resulting from the Proctor compaction test. It can be considered that the shearing behaviour was governed by the inclusion of waste straw when %waste straw < 0.75 and that it, in turn, was governed by the densification effect when %waste straw > 0.75. %waste straw > 0.75 aggravated the densification effect and inclusions of waste straw more than 1% were, therefore, neglected in this work. Marginal values, namely, 0.3 and 0.75, should not be chosen as the optimal %waste straw towards ensuring the effectiveness of reinforcement. Further, plastic and liquid limits for the loess distributed national wide varies in a range of 16.6–19.3 and 25.2–31.8 respectively, according to Tian et al. (1987). On the other hand, the optimal ω obtained from the standard Proctor test result (not shown here) measured at 20%. In this context, %waste straw = 0.7 and ω = 18% were fed back to the preparation of specimens subjected to stress-controlled direct shear.

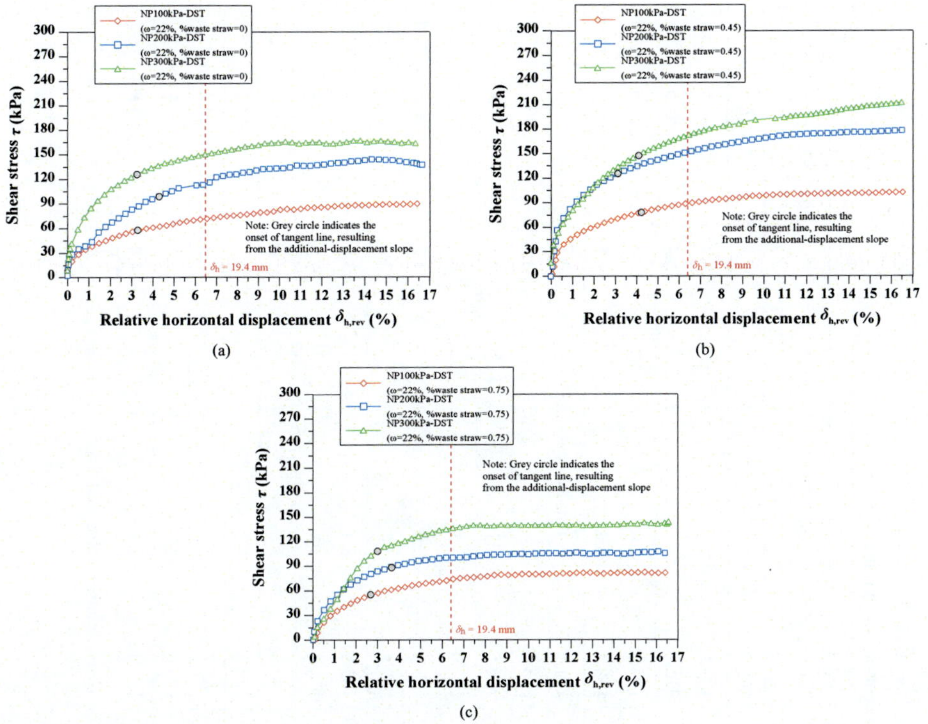

Fig. 9. Shear stress τ versus relative horizontal displacement $\delta_{h,rev}$ relation for specimens at $\omega = 22\%$ with (a) %waste straw = 0, (b) %waste straw = 0.45, and (c) %waste straw = 0.75 under the normal pressures of 100 kPa, 200 kPa, and 300 kPa

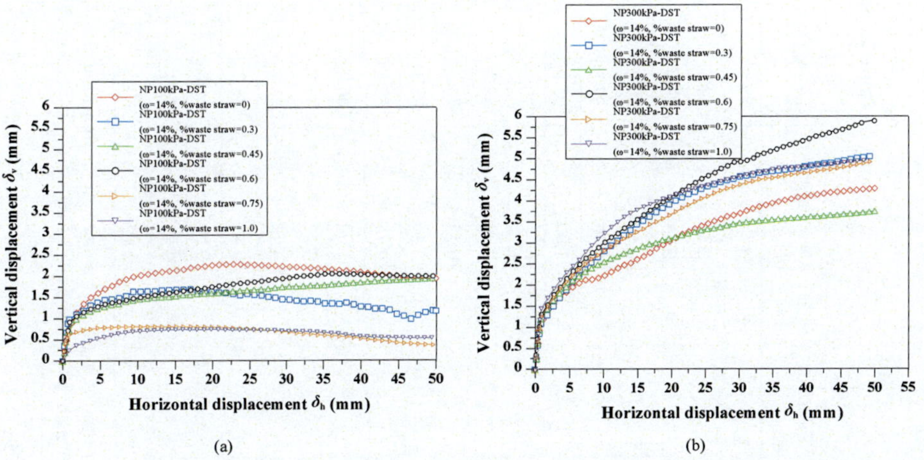

Fig. 10. Vertical displacement δ_v versus horizontal displacement δ_h for specimens at $\omega = 14\%$ under (a) $\sigma_n = 100$ kPa and (b) $\sigma_n = 300$ kPa

Table 2. Summary of the shear strength parameters

Specimen	Shear box (mm)	ω (%)	%waste straw (%)	$\tau_{ult-100}$ (kPa)	$\tau_{ult-200}$ (kPa)	$\tau_{ult-300}$ (kPa)	$\delta_{h,rev-100}$ (%)	$\delta_{h,rev-200}$ (%)	$\delta_{h,rev-300}$ (%)	c (kPa)	ϕ (°)
Lantian loess	305	14	0	77.7	107.8	139.9	4.3	4.4	3.1	46.2	17.3
			0.30	88.3	130.4	173.6	3.2	4.5	4.1	45.5	23.1
			0.45	103.9	167.8	215.0	3.2	3.2	5.4	51.0	29.1
			0.60	80.2	132.4	194.2	2.6	4.7	5.1	21.6	29.7
			0.75	97.3	115.7	158.1	3.4	3.4	3.8	62.9	16.9
			1.00	105.9	153.3	191.9	3.3	4.0	4.7	64.4	23.3
		18	0	70.9	106.4	145.0	2.4	3.5	3.9	33.4	20.3
			0.30	89.2	137.5	166.5	3.1	4.1	4.8	53.8	21.1
			0.45	73.3	120.8	157.0	3.1	2.8	4.9	33.3	22.7
			0.60	85.4	116.0	168.4	5.0	4.8	3.7	40.3	22.5
			0.75	82.2	103.9	137.8	4.8	3.1	3.0	52.4	15.5
			1.00	66.9	119.2	163.5	2.4	3.4	3.6	19.9	25.8
		22	0	57.7	100.4	126.2	3.3	4.2	3.3	26.3	18.9
			0.30	56.7	105.8	141.2	3.8	4.6	3.6	16.7	22.9
			0.45	76.2	124.5	149.4	4.2	3.1	4.2	43.5	20.1
			0.60	65.8	122.1	162.3	3.3	5.3	4.7	20.2	25.8
			0.75	56.1	88.3	103.6	2.7	3.7	2.7	35.2	13.4
			1.00	80.3	110.8	139.8	4.4	4.3	4.0	50.8	16.6

Note: ω = water content; $\tau_{ult-100}$ = ultimate shear strength under σ_n = 100 kPa; $\delta_{h,rev-100}$ = relative horizontal displacement to $\tau_{ult-100}$; c = cohesion associated with ultimate shear strength; ϕ = friction angle associated with ultimate shear strength.

Table 3. Summary of the large-displacement shear strength parameters

Specimen	Shear box (mm)	ω (%)	%waste straw (%)	$\tau_{\text{ld-100}}$ (kPa)	$\tau_{\text{ld-200}}$ (kPa)	$\tau_{\text{ld-300}}$ (kPa)	δh, rev (%)	c_{ld} (kPa)	ϕ_{ld} (°)
Lantian loess	305	14	0	107.5	174.5	240.5	16.4	41.2	33.6
			0.30	121.1	196.4	274.5	16.4	43.9	37.5
			0.45	136.7	229.7	278.0	16.4	73.4	35.3
			0.60	135.3	194.8	271.1	16.4	64.6	34.2
			0.75	110.5	192.3	274.2	16.4	28.6	39.3
			1.00	147.7	206.3	280.7	16.4	78.6	33.6
		18	0	104.9	172.8	235.1	16.4	40.7	33.1
			0.30	121.2	195.6	246.7	16.4	62.3	32.1
			0.45	101.5	178.6	211.2	16.4	54.0	28.8
			0.60	113.1	158.7	253.5	16.4	34.7	35.1
			0.75	102.3	151.6	215.7	16.4	43.1	29.6
			1.00	115.3	193.4	267.8	16.4	39.7	37.3
		22	0	88.1	136.5	162.6	16.4	54.6	20.4
			0.30	95.2	149.1	189.1	16.4	50.5	25.2
			0.45	102.4	176.8	211.2	16.4	54.7	28.5
			0.60	89.9	137.3	181.0	16.4	44.9	24.5
			0.75	80.9	105.1	141.3	16.4	48.7	16.8
			1.00	102.3	151.6	170.2	16.4	73.4	18.7

Note: ω = water content; $\tau_{\text{ld-100}}$ = large-displacement shear strength under $\sigma_n = 100$ kPa; c_{ld} = cohesion associated with large-displacement shear strength; ϕ_{ld} = friction angle associated with large-displacement shear strength.

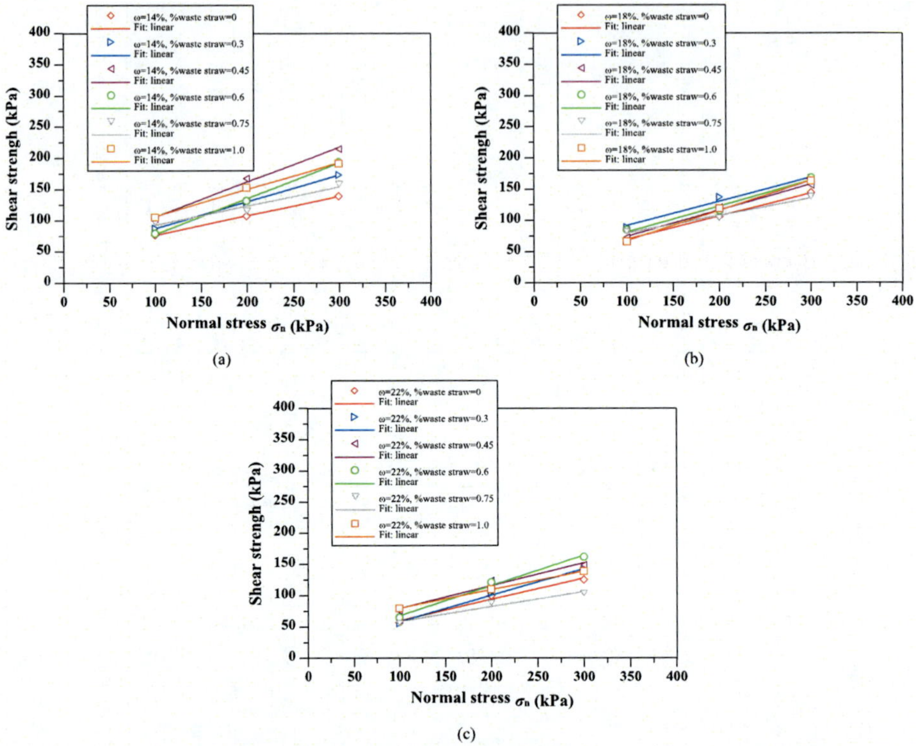

Fig. 11. Shear strength envelopes for specimens at (a) ω = 14%, (b) ω = 18%, and (c) ω = 22%

3.3 Shear Behaviour in Stress-Controlled Direct Shear Tests

Stress-controlled direct shear tests on Lantian loess specimens at $\omega = 18\%$, with %waste straw = 0.7, were conducted. In order to highlight the effect of waste straw inclusions, the results were compared to those from Lantian specimens at $\omega = 18\%$, without waste straw inclusion. Temporal relations of horizontal displacement δ_h, shear deformation rate, and vertical displacement δ_v for the specimens with %waste straw = 0 and 0.75 are depicted in Figs. 12 and 13, respectively. Vertical dash lines in Figs. 12 and 13 identify an increase in τ of 25 kPa and the number adjacent to each line stands for the total τ. On the whole, an increase in δ_h and δ_v and high shear deformation rate were observed following each shear load increment. The specimen without waste straw inclusion experienced δ_h and δ_v increasing to 0.3 mm and 0.2 mm respectively when subjected to τ = 25 kPa and thereafter increasing to 11.2 mm and 1.1 mm respectively when subjected to τ = 50 kPa. The specimen experienced δ_h increasing rapidly to 50 mm, with a shear deformation rate increasing to 21.6 mm/min, when subjected to $\tau_f = 75$ kPa, indicating that the failure of specimen was attained. While the specimen experienced δ_v increasing ultimately to 3.2 mm. In contrast, the specimen with %waste straw = 0.75 experienced δ_h and δ_v increasing to 12.2 mm and 1.9 mm respectively without showing any signs of failure when subjected to τ = 75 kPa. The specimen reaching failure at τ_f = 125 kPa was characterised by δ_h increasing rapidly to 46 mm, with a shear deformation rate

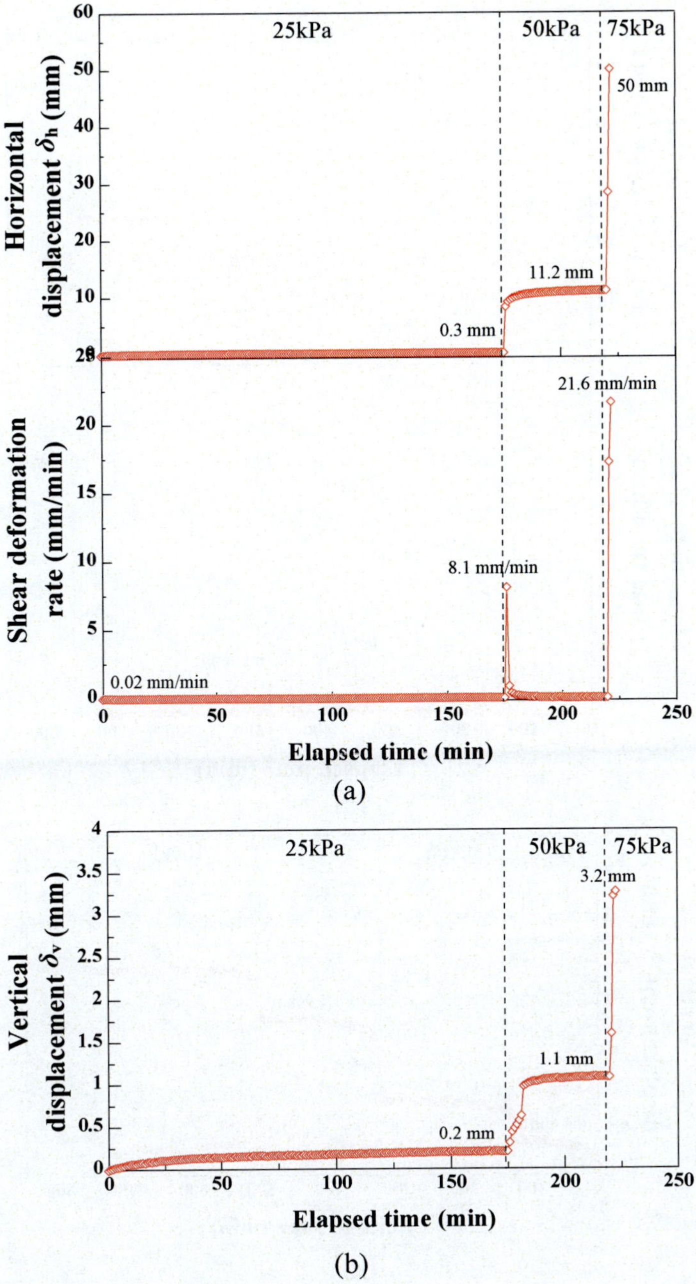

Fig. 12. Temporal relations of (a) horizontal displacement δ_h and deformation rate and (b) vertical displacement δ_v for the specimen with %waste straw = 0

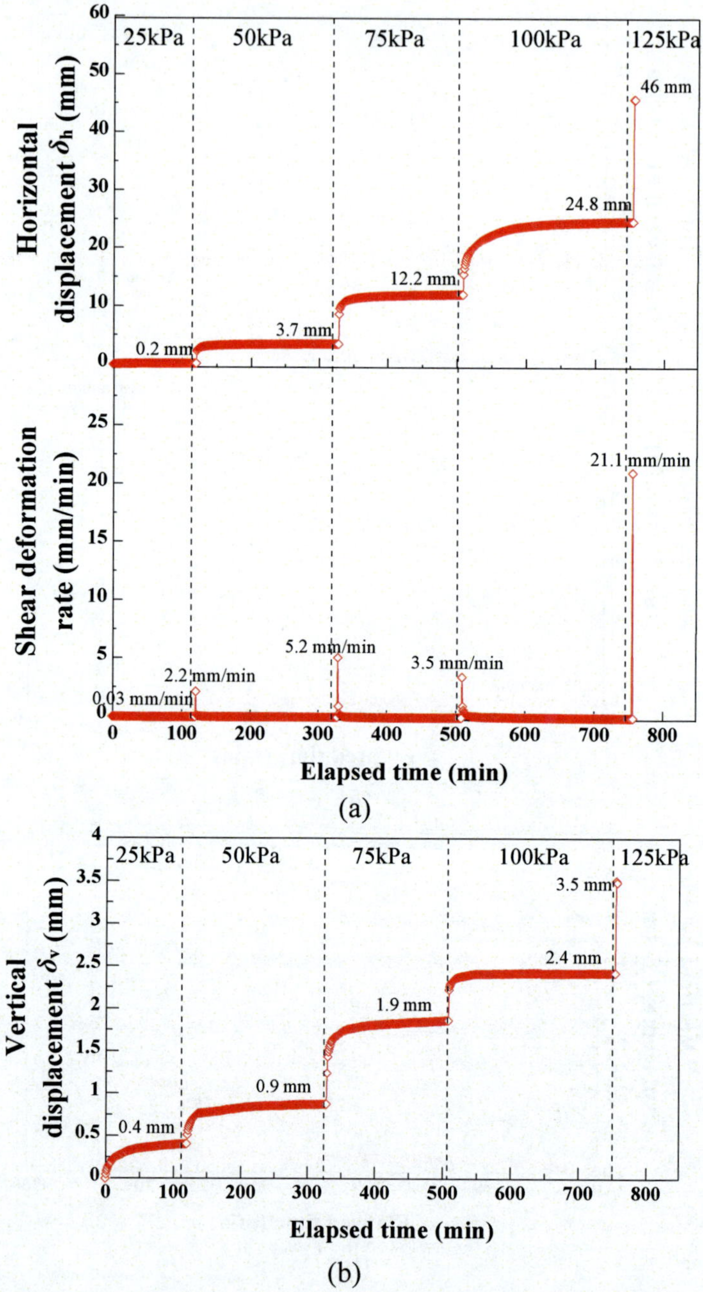

Fig. 13. Temporal relations of (a) horizontal displacement δ_h and deformation rate and (b) vertical displacement δ_v for the specimen with %waste straw = 0.75

increasing to 21.1 mm/min. δ_v associated with τ_f measured at 3.5 mm. The results from the stress-controlled direct shear tests are summarised in Table 4.

Table 4. Summary of the stress-controlled direct shear tests

Specimen	Shear box (mm)	ω (%)	%waste straw (%)	σ_n (kPa)	Tf (kPa)	ϕ_s (°)
Lantian loess	305	18	0	95	75	38.3
			0.7	95	125	52.8

Note: ω = water content; σ_n = normal pressure on specimen; τ_f = shear stress at failure; ϕ_s = secant friction angle = $\tan^{-1}(\tau_f/\sigma_n)$.

4 Discussion

4.1 Effect of Agricultural Waste Straw Inclusions

Preventing instability of loess in hilly-gullied regions of northwest China is deemed to be of great importance towards protecting loessial environment and reducing the potential of landslide hazard. Agricultural waste straws reinforce the tested loess by transferring shear stresses that develop in the soil matrix into tensile resistance in the waste straw inclusions through interface friction along the length of imbedded waste straws. A simple force equilibrium model was introduced here and effective in accentuating the effect of waste straw inclusions and predicting thc amount of shear strength increase. As subjected to shear, waste straws deform as depicted in Fig. 14. In case waste straws oriented initially at an acute angle (<90°) to the shear plane or in the direction of maximum principle tensile strain lead to the greatest increase in shear strength. In contrast, an oblique orientation with the shear plane (>90°) can cause shear strength decrease because the waste straws actually go into compression rather than tension. The simple, perpendicular model, therefore, can provide a very useful simulation, although waste straws may have many orientations and are unlikely to be oriented perpendicularly to the shear plane.

Based on this perpendicular model, the increase in shear strength $\Delta\tau$, induced by the waste straw inclusions, can be expressed as:

$$\Delta\tau = t_{ws}(\sin\theta + \cos\theta\tan\phi) \tag{1}$$

$$\theta = \tan^{-1}(\delta_h/t) \tag{2}$$

where $\Delta\tau$ = shear strength increase, θ = angle of shear distortion, δ_h = horizontal displacement, t = shear zone thickness, and t_{ws} = mobilised tensile stress of waste straws per unit area of soil. $\Delta\tau$ was calculated by considering the following conditions: δ_h = 12 mm ($\delta_{h,rev}$ = 4%) and shear zone thickness t = 1 mm. Thus, when $\theta \approx 90°$, Eq. (1) can be rewritten as:

$$\Delta\tau = t_{ws} \tag{3}$$

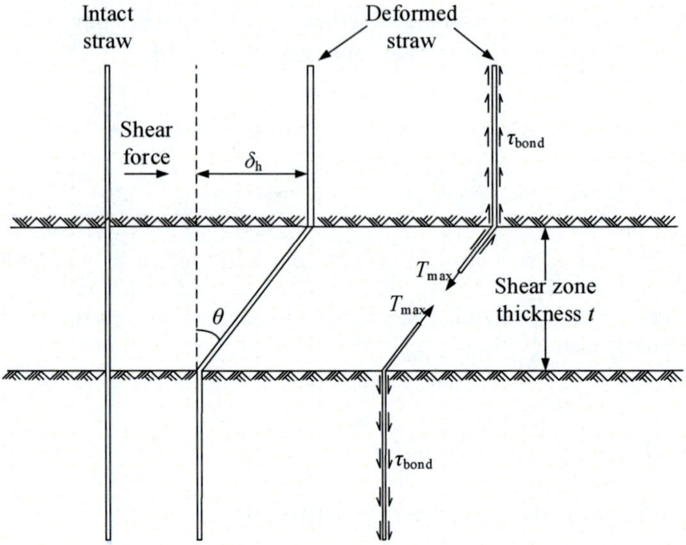

Fig. 14. Schematic diagram of the perpendicular waste straw reinforcement model

The results indicate that except a small number of cases showing shear strength worse than those without waste straw inclusion, the increase in c, induced by the waste straw inclusions, ranged from 4.8 kPa to 24.5 kPa, whereas the increase in ϕ ranged from 0.8° to 12.4° (Table 2). These results corresponded to the increase in c ranging from 14.4% to 73.4% and the increase in ϕ ranging from 3.9% to 61% compared to the c and ϕ values from the specimen at $\omega = 18\%$ without waste straw inclusion. To verify the observations in the present study, several previous studies were, therefore, chosen to be compared in terms of the effect of reinforcing material inclusion. Comparisons of the present study and the previous studies are tabulated in Table 5. Tan et al. (2019) investigated the shear strength of the fallow soil using direct shear method. The increase in c for the fallow soil, reinforced with root and geomat, ranged from 10.9 kPa to 23.2 kPa in comparison with the increase in c being 3.7 kPa and 6.6 kPa for the fallow soil reinforced with root and geomat respectively. Sahin Zaimoglu (2010) studied a natural fine grained soil, classed as high plasticity silt (MH) and reinforced with 0.25–1% polypropylene fibers, using direct shear method. The results indicated that c increased with the inclusion of fibers, whereas ϕ was not affected significantly by the fiber reinforcement. To evaluate the effects of nanoclay stabilisation, varying inclusions of nanoclay ranging from 0.2% to 3% by mass were added to the wind-blown loess from Golestan Province, Iran (Tabarsa et al. 2018). Specimens with and without nanoclay were tested under unconsolidated undrained triaxial compression conditions for confining pressures of 100 kPa, 200 kPa, and 300 kPa. The results indicated that %nanoclay = 1, 2, and 3 caused a significant increase in c and a small reduction in ϕ. Sivakumar Babu and Raja Jaladurgam (2014) explored the potential use of fly ash mixed with plastic waste in civil engineering applications and found that with increase of %plastic waste, ϕ for 38 × 76 mm specimens increased from 25° to 26° and c increasing from 19.5 kPa to 34 kPa. ϕ for 50 × 100 mm

Table 5. Comparison of the present study and previous studies

Soil type	USCS symbol	Testing method	Reinforcing material content (%)	Normal pressure σ_n or confining pressure σ_3 (kPa)	Increase in c (kPa)	Increase in ϕ (deg)
Root-geomat reinforced silty clay (Tan et al. 2019)	CL	Displacement-controlled direct shear	0.23–0.58 (root) 0.41–1.75 (root + geomat)	25, 50, and 100	10.9–23.2 (root + geomat)	NA
					3.7 (root)	
					6.6 (geomat)	
Polypropylene fiber reinforced high plasticity silt (Sahin Zaimoglu and Yetimoglu 2012)	MH	Displacement-controlled direct shear	0.25–1 (randomly distributed polypropylene fiber)	50, 100, and 200	23–86	10–14
Nanoclay reinforced loess (Tabarsa et al. 2018)	CL-ML	Unconsolidated undrained triaxial compression	0.2–3 (nanoclay)	100, 200, and 300	10.8–28.7	NA
Plastic waste reinforced fly ash (Sivakumar Babu and Raja Jaladurgam 2014)	NA	Consolidated undrained triaxial compression	0.5–1 (plastic waste)	50, 100, and 150	14.5 (small specimen)	1 (small specimen)
					38 (large specimen)	1 (large specimen)
Agricultural waste straw reinforced loess (this study)	ML	Displacement-controlled and stress-controlled direct shear	0.3–1 (waste straw)	100, 200, and 300 (displacement-controlled direct shear)	4.8–24.5	0.8–12.4
				95 (stress-controlled direct shear)		

Note: NA = not applicable.

specimens increased from 27° to 28° and c increased from 17 kPa to 55 kPa. The above relations of increase in shear strength and different %reinforcing material were regressed using logarithmic function respectively as depicted in Fig. 15. On the whole, increase in τ with increasing %reinforcing material was observed, with a more pronounced effect at lower contents, and increase in c contributed the most to the increase in τ of reinforced specimen. In summary, the previous studies still provide strong support for the increases in τ with increasing %reinforcing material identified in the present study.

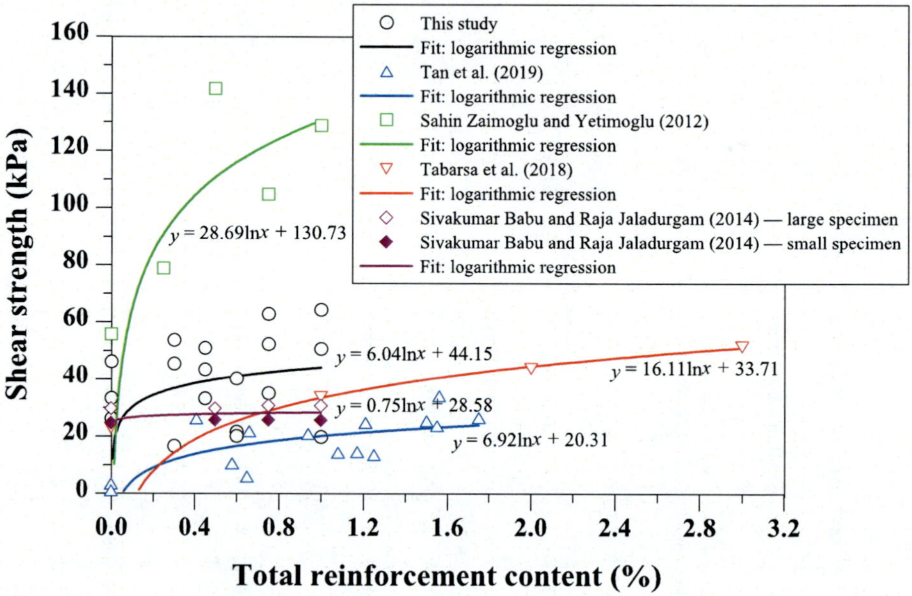

Fig. 15. Relations of shear strength versus total reinforcement content for the present study and previous studies

4.2 Effect of Agricultural Waste Straw Content

To evaluate the effect of waste straw content, a fraction of loess cross-section occupied by agricultural waste straw A_{WS}/A, namely, straw area ratio, was introduced into the perpendicular model. The 'straw area ratio' can be determined by counting waste straws by size within a given loess area as:

$$A_{WS}/A = (\Sigma na)/A \tag{4}$$

where n = number of waste straws within a given loess cross-section A, and a = mean cross-sectional area of waste straws. The mobilised tensile stress t_{WS} in fact depends upon the amount of waste straw elongation and fixity of the waste straws in the soil matrix. Full mobilization of t_{WS} can only occur if the waste straws elongate sufficiently and if imbedded waste straws are prevented from slipping or pulling out. Accordingly, there are three behaviours likely to occur during shearing of a loess-waste straw mixture, namely, waste straws break, stretch, and slippage. In our case the waste straws are very short (i.e. 40 mm long), randomly distributed, and unconstrained, they will tend to slip or pull out when the loess-waste straw mixture is sheared. At initial slippage, the maximum tension in a waste straw T_{max} is given by:

$$T_{max} = 2\tau_{bond}L/D \tag{5}$$

$$\tau_{bond} = h\gamma(1 - \sin\phi)\mu\tan\phi \tag{6}$$

where τ_{bond} = interface friction stress between waste straw and loess, L = waste straw length, D = waste straw diameter, h = depth below ground surface, γ = waste straw density, and μ = frictional coefficient. The mobilised tensile stress per unit area of loess in this case is given by:

$$t_{ws} = (2\tau_{bond}L/D)(\Sigma na)/A \tag{7}$$

Substituting Eq. (7) into Eq. (1) gives:

$$\Delta\tau = (\pi\tau_{bond}/2A)\,[\sin\theta + \cos\theta\tan\phi]\Sigma nLD \tag{8}$$

To investigate the effect of τ_{bond} on the shear strength increase and how μ affects τ_{bond}, τ_{bond} and μ were back-analysed under the following conditions: L = 40 mm, D = 1 mm, and γ = 2.3 kN/m^3. Relation of increase in c versus τ_{bond} is depicted in Fig. 16. It is evident that increase in c increased logarithmically with increasing τ_{bond} and μ for the loess and waste straw interface ranged from 0.07 to 0.24 in comparison with μ of 0.7–0.9 for the soil and wood interface (Gray and Barker 2004). The rough texture and kinky shape of plant roots were deemed as the main cause to lead to the higher μ.

Given the unit weight of straw γ = 2.3 kN/m^3, %waste straw was converted to the equivalent straw area ratio to evaluate the effect of waste straw content. Relations of increase in c versus straw area ratio from the measured results and predictions considering different waste straw diameters D ranging from 2 mm to 5 mm are depicted in Fig. 17. The measured increase in c increased in a nonlinear manner with increasing straw area ratio. It can also be seen that the measured results resulting from D = 1 mm were in reasonable agreement with the prediction resulting from D = 2 mm and the prediction resulting from D = 1 mm should lie above the measured results, meaning that the perpendicular model produced an overestimation. This overestimation could be most likely due to the discrepancy between the orientation of waste straws assumed in the perpendicular model and that of reality. The predictions also reveal that increases in c varied inversely with D. Accordingly, at equal straw area ratios numerous small diameter straws will be more effective in accentuating the effect of waste straw inclusions than a few large diameter straws. On the other hand, nonlinearity in the measured increase in c appeared not to be captured by the perpendicular model. Relations of increase in c versus straw area ratio from the measured results and predictions considering different straw lengths L ranging from 5 mm to 30 mm are also depicted in Fig. 17. It can be observed from Fig. 17 that the increment of increases, induced by considering different straw lengths L, was larger

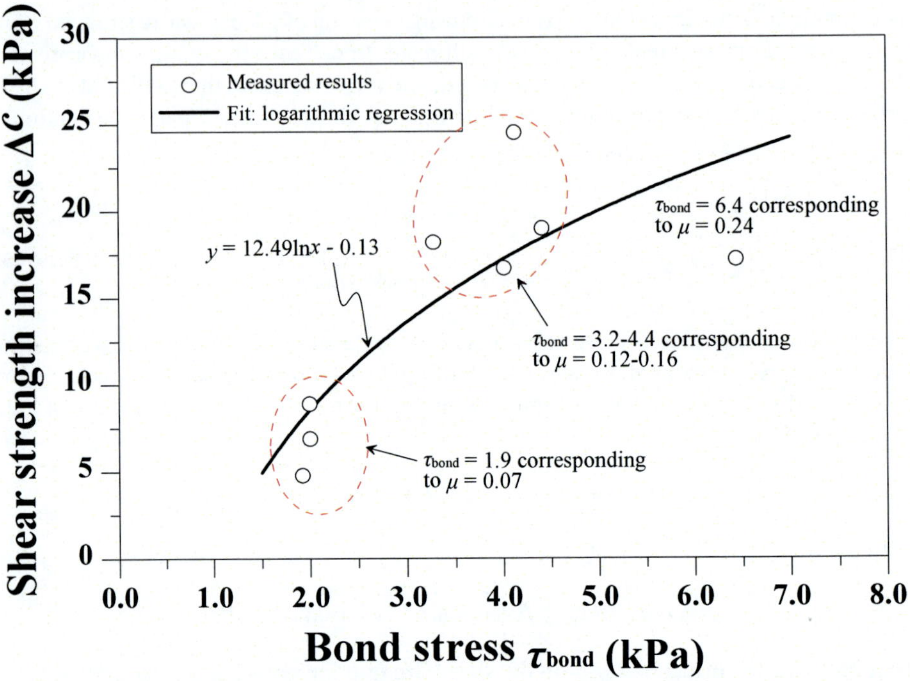

Fig. 16. Relation of shear strength increase versus bond stress

than that induced by considering different straw diameters *D* and increases in *c* were in proportional relation with *L*.

To better understand the intrinsic contribution of straw inclusion to the increase in shear strength, dilation behaviour may be of worthy for investigation, and it is closely related to the rotation and breakage of coarse-grained particles. In this work, the dilation of aggregates mainly derived from the rotation of inter-particle waste straws or could be suppressed by the deformation of loess. The dilation procedure can be evaluated using the dilation rate, as defined in Eq. (9).

$$\tan\psi = d\delta_v / d\delta_h \tag{9}$$

where $d\delta_v$ is the vertical deformation increment and $d\delta_h$ is the shear deformation increment. Figure 18 shows the relationship between the shear stress and dilation rate versus horizontal displacement under the normal pressures of 100 kPa, 200 kPa, and 300 kPa. The value of ψ are compared to the value of $\phi_{ld} - \phi_p$. Table 6 lists ϕ_p, ϕ_{ld} and ψ values against different water contents and %waste straws. It is substantiated by the test results that ϕ_p and ϕ_{ld} decreased notably with increasing normal stress. The value of $\phi_{ld} - \phi_p$ varied within the range of 2.5–14.8°, while the value of ψ varied within the range of 0.5–17.2°. It can be seen from Fig. 19 that the data varied less than 3°, while in some case it was not towards verifying the applicability of this approach. The contribution of the dilation to the increase in shear strength can be effectively interpreted either using ψ or $\phi_{ld} - \phi_p$.

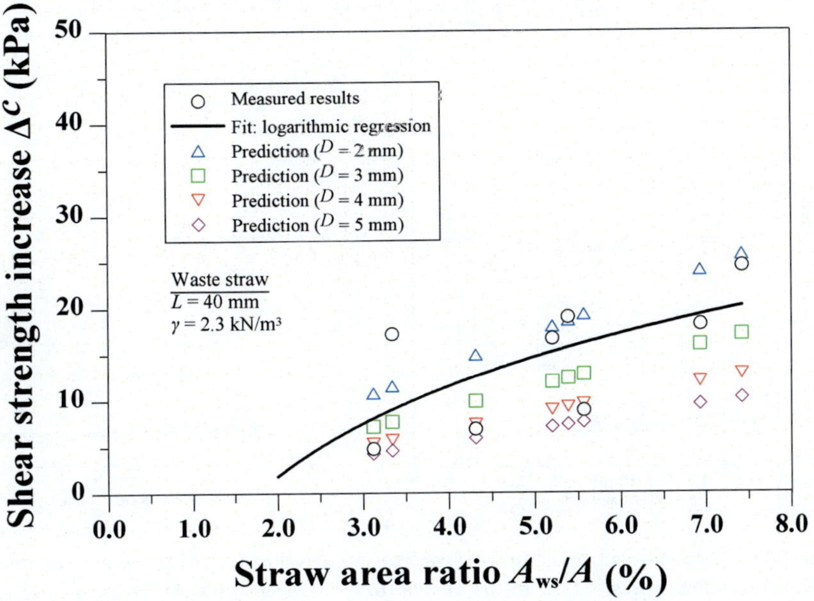

(a)

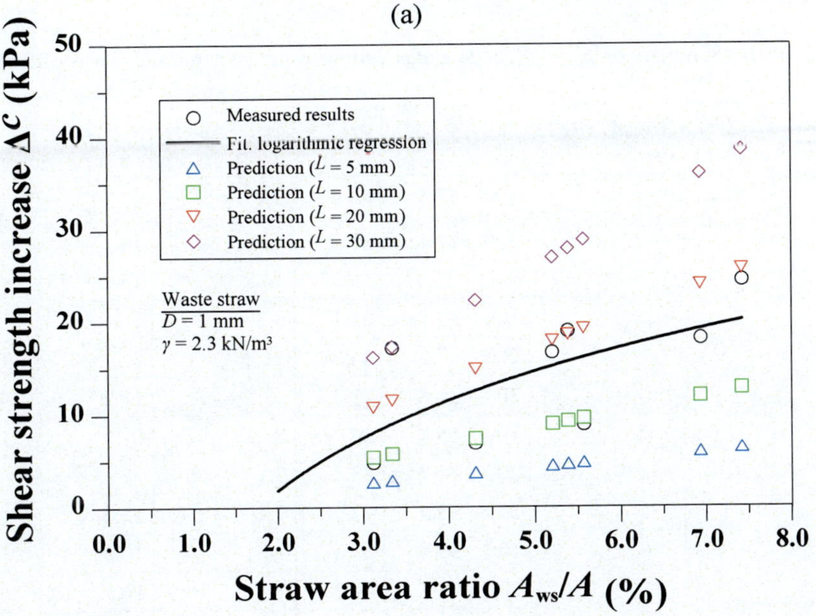

(b)

Fig. 17. Relations of shear strength increase versus straw area ratio from the measured results and predictions considering: (a) different straw diameters and (b) different straw lengths

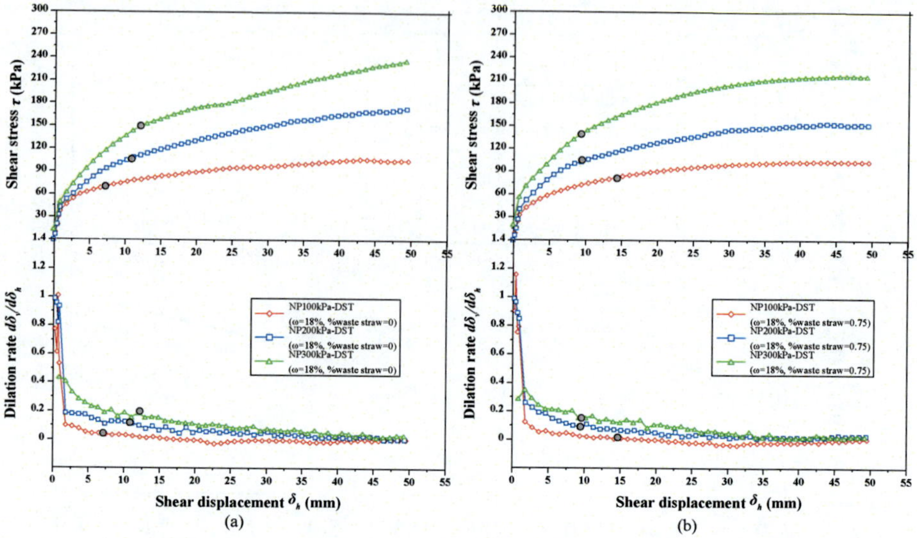

Fig. 18. Displacement-controlled direct shear and stress-controlled direct shear comparison for: (a) specimens without waste straw inclusion and (b) specimens with waste straw inclusions

Table 6. Summary of $\phi_{ld} - \phi_p$ and against various water contents and %waste straw

(%)	%waste straw (%)	σ_n	ϕ_p (°)	ϕ_{ld} (°)	$\phi_{ld} - \phi_p$ (°)	ψ (°)
14	0	100	37.8	46.9	9.1	9
		200	28.1	41	12.9	14.1
		300	25.2	39	13.8	13.2
	0.3	100	41.8	51.1	9.3	4.3
		200	33.2	44.5	11.3	9.7
		300	29.8	42.3	12.5	13.2
	0.45	100	45.8	54	8.2	6
		200	39.7	48.8	9.1	12
		300	35.5	42.9	7.4	12.6
	0.6	100	39	53.4	14.4	10.6
		200	33.3	44.3	11	14.5
		300	32.6	42.1	9.5	14.4
	0.75	100	44.6	48	3.4	1.6
		200	29.9	43.7	13.8	13.9
		300	28	42.5	14.5	13.4

(*continued*)

Table 6. (*continued*)

(%)	%waste straw (%)	σ_n	ϕ_p (°)	ϕ_{ld} (°)	$\phi_{ld} - \phi_p$ (°)	ψ (°)
	1.0	100	47.2	56.2	9	4.5
		200	37.4	45.8	8.4	6.4
		300	32.5	43.2	10.7	12.9
18	0	100	35.5	46.4	10.9	9.2
		200	27.7	40.9	13.2	14.2
		300	26.3	38.2	11.9	10.7
	0.3	100	41.7	50.8	9.1	7.2
		200	34.4	44.2	9.8	11.6
		300	30.3	40.1	9.8	7.5
	0.45	100	36.1	45.7	9.6	7.6
		200	30.9	41.7	10.8	13
		300	27.7	35.2	7.5	5.7
	0.6	100	40.2	48.9	8.7	14.2
		200	30.4	38.3	7.9	7
		300	29.4	40.3	10.9	8.9
	0.75	100	39.4	45.9	6.5	0.5
		200	27.3	37.2	9.9	12.5
		300	24.8	35.8	11	8.8
	1.0	100	33.8	48.6	14.8	9.7
		200	30.9	44	13.1	13.6
		300	28.7	41.9	13.2	17.2
22	0	100	30.3	42	11.7	10.5
		200	23.6	34.5	10.9	13.7
		300	23	28.6	5.6	7.7
	0.3	100	29.8	43.7	13.9	13.9
		200	28.1	36.7	8.6	7.6
		300	25.3	32.3	7	9.6
	0.45	100	35.5	40.5	5	4.4
		200	25.2	31.1	5.9	6.5
		300	21.9	244	2.5	3.9
	0.6	100	34.1	42.4	8.3	5.5
		200	31	34.7	3.7	4.4
		300	28.6	31.2	2.6	4.2

(*continued*)

Table 6. (*continued*)

(%)	%waste straw (%)	σ_n	ϕ_p (°)	ϕ_{ld} (°)	$\phi_{ld} - \phi_p$ (°)	ψ (°)
	0.75	100	28.9	39.3	10.4	12
		200	24.1	27.9	3.8	6.2
		300	19	25.7	6.7	8
	1.0	100	38.8	45.1	6.3	9
		200	29.2	37.5	8.3	6.8
		300	25.3	29.7	4.4	4.2

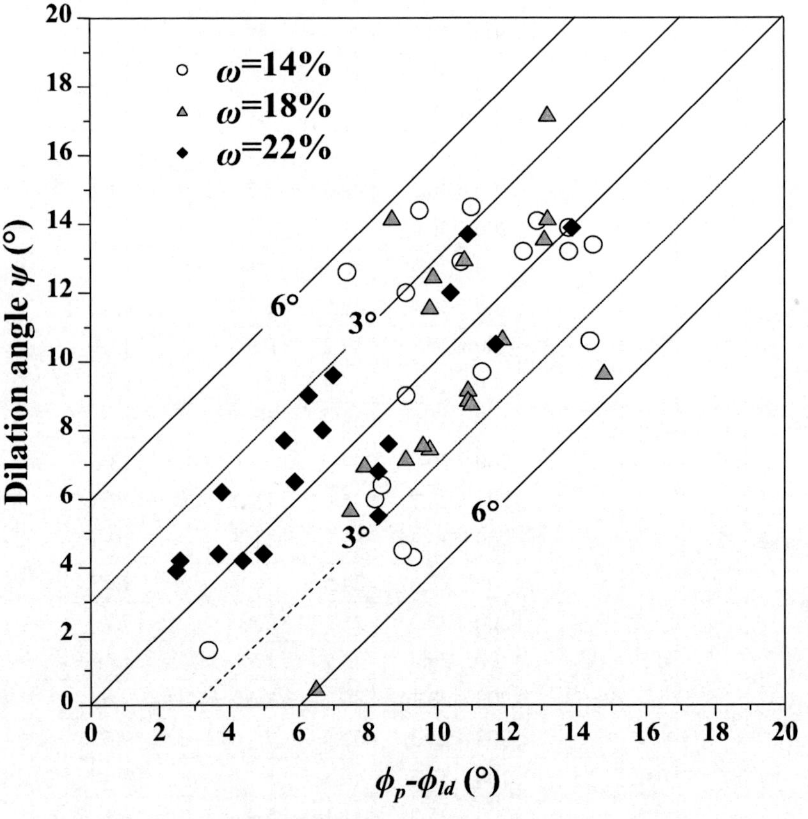

Fig. 19. Comparison of $\phi_{\mathrm{ld}} - \phi_{\mathrm{p}}$ and ψ

4.3 Displacement-Controlled Direct Shear and Stress-Controlled Direct Shear Comparison

During the stress-controlled direct shear tests, δ_h increased with increasing τ for both specimens, as shown in Figs. 12 and 13. δ_h for the specimen without waste straw inclusion continued to increase until a point identified the start of failure. This point was characterised as (i) a change in curvature of the δ_h versus t relation from positive to negative and (ii) a rapid increase in δ_h. The shear resistance reached a maximum, which represents shear strength, immediately before failure. However, τ could not be resisted as the shear resistance was reduced; therefore, the unreinforced specimen experienced a rapid accumulation of δ_h after the onset of failure. δ_h and δ_v finally reached 50 mm and 3.2 mm respectively. In contrast, δ_h for the specimen with %waste straw = 0.7 increased to 12.2 mm when subjected to $\tau = 75$ kPa and continued to increase until a starting point of failure when subjected to $\tau = 125$ kPa. Prior to failure, the shear resistance reached a maximum (i.e. shear strength). τ could not be resisted due to the reduced shear resistance leading to a rapid accumulation of δ_h following the commencement of failure. δ_h and δ_v for the reinforced specimen ultimately measured at 46 mm and 3.5 mm respectively. It can be realised that compared with the specimen without waste straw inclusion, the effect of waste straw inclusions was well manifested by transferring τ that develops in the soil matrix into tensile resistance in the waste straws through interface friction along their length and deemed as the main cause to lead to the higher shear strength. Relations between τ_f (or τ_{ult}) and σ_n from displacement-controlled direct shear tests undertaken on specimens without waste straw inclusion and from stress-controlled direct shear tests undertaken on same specimens are depicted in Fig. 18. It can be observed that for the specimens without waste straw inclusion, τ_{ult} from the displacement-controlled direct shear test under $\sigma_n = 100$ kPa showed good correspondence with τ_f from the stress-controlled direct shear test under similar σ_n (i.e. 95 kPa). The τ_f (or τ_{ult})$-\sigma_n$ data were regressed using logarithmic function. On the other hand, τ_{ult} from the displacement-controlled direct shear test on the specimen with %waste straw = 0.6 under $\sigma_n = 100$ kPa also showed good correspondence with τ_f from the stress-controlled direct shear test on the specimen with %waste straw = 0.7. The reason why τ_{ult} from the specimen with %waste straw = 0.75 to deviate further from τ_f was most likely because of the effect of densification, induced when %waste straw equal to or in excess of 0.75. To sum it up, the stress state at failure in a stress-controlled direct shear test and a displacement-controlled direct shear test for same specimen under identical testing conditions showed good correspondence to each other irrespective of testing method and approximated a single failure envelope (Fig. 20).

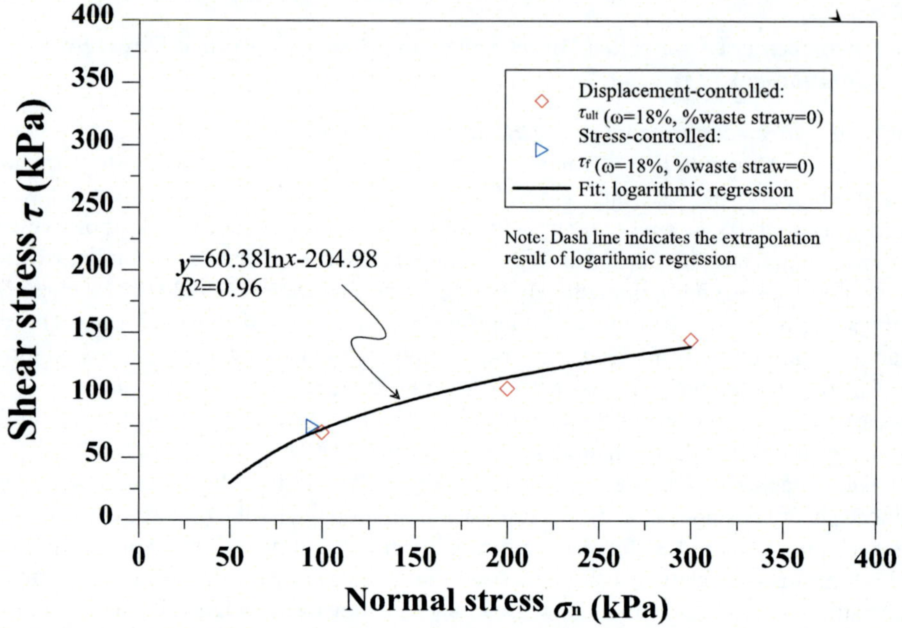

(a)

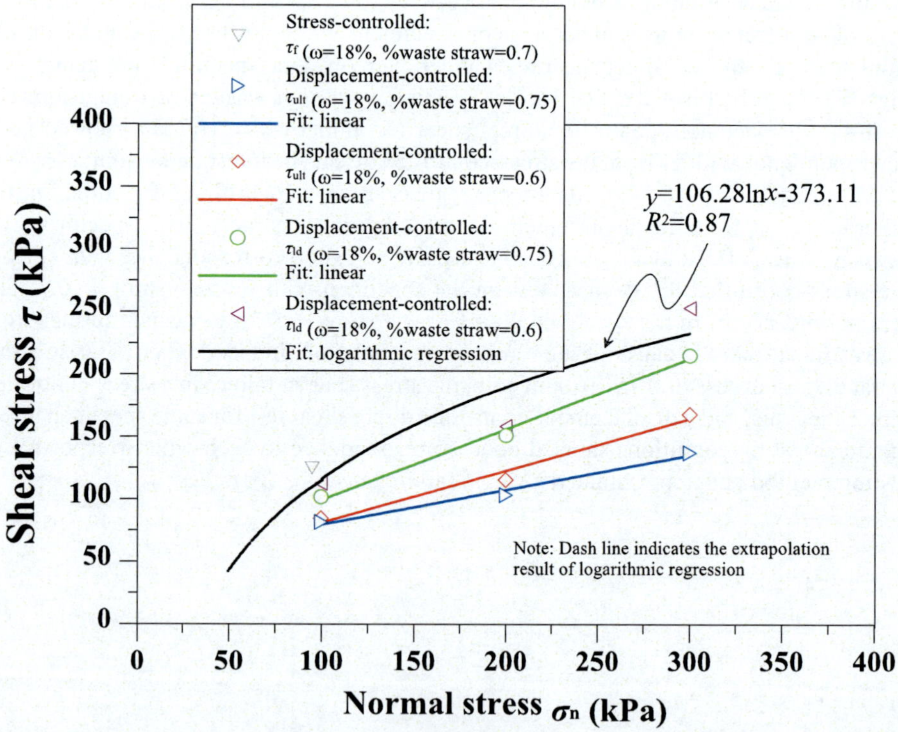

(b)

Fig. 20. Variation of shear stress and dilation rate against shear displacement: (a) %waste straw = 0 and (b) %waste straw = 0.75

5 Conclusions

This study performed the displacement-controlled direct shear tests on Lantian loess with and without waste straw inclusion under three normal pressures to document the baseline strength parameters. Further, to investigate the temporal relation of horizontal and vertical displacement under given shear stresses, the stress-controlled direct shear tests on the same soil with and without waste straw inclusion were conducted. Based on the results and discussion, some main conclusions can be drawn as follows:

(1) Prior to the displacement-controlled direct shear tests, testing method was validated using six repeatability tests. The relations of shear stress versus relative horizontal displacement were classed into three types for which τ_{ult} were defined by different manners. Further, the baseline shear strength parameters intervened the preparation of specimens tested in the stress-controlled direct shear tests.

(2) The increase in shear strength, induced by the waste straw inclusions, was measured and predicted using the perpendicular model. The effect of waste straw inclusions was well manifested by transferring shear stress that develops in the soil matrix into tensile resistance in the waste straws through interface friction along their length. The relations between τ_f (or τ_{ult}) and σ_n from the displacement-controlled direct shear test and from the stress-controlled direct shear test showed good correspondence to each other irrespective of testing method. Further, their results approximated a single failure envelope.

(3) This study explored an exciting potential use of the loess-waste straw mixture as an alternative backfill material for agricultural solid waste functionalisation and improvement of loess stability against shallow slope failure. Its horizon of practical applications can be widened with the superior shearing behaivour identified in this study.

6 Data Availability

The experimental data used to support the findings of this research work are included in the article.

7 Conflicts of Interest

The authors declare that they have no known competing financial interests or personal relationships that could have appeared to influence the work reported in this paper.

Acknowledgements. This study would not have been possible without financial supports from the Special Fund for Basic Scientific Research of Central Colleges, Chang'an University, under Grant no. 300102269502.

Author Contribution. This paper represents a result of collaborative teamwork. Mr. Zhong-Fei Xue and Miss Lin Wang performed soil sampling and a series of laboratory tests; Dr. Wen-Chieh Cheng and Dr. Jian Xu guided the laboratory tests and prepared the original manuscript. The four authors contributed equally to this work.

References

Crosta, G.B., Imposimato, S., Roddeman, D.: Numerical modelling of entrainment/deposition in rock and debris-avalanches. Eng. Geol. **109**, 135–145 (2008)

Crosta, G.B., Imposimato, S., Roddeman, D., Frattini, P.: On controls of flow-like landslide evolution by an erodible layer. In: Margottini, C., et al. (eds.) Landslide Science and Practice. Springer, Heidelberg (2013). https://doi.org/10.1007/978-3-642-31310-3_35

Cheng, W.C., Ni, J.C., Arulrajah, A., Huang, H.W.: A simple approach for characterising tunnel bore conditions based upon pipe-jacking data. Tunn. Undergr. Space Technol. **71**, 494–504 (2018)

Cheng, W.C., Song, Z.P., Tian, W., Wang, Z.F.: Shield tunnel uplift and deformation characterisation: a case study from Zhangzhou Metro. Tunn. Undergr. Space Technol. **79**, 83–95 (2018)

Cheng, W.C., Ni, J.C., Shen, S.L., Wang, Z.F.: Modeling of permeation and fracturing grouting in sand: laboratory investigations. J. Test. Eval. **46**(5), 2067–2082 (2018)

Cheng, W., Xue, Z., Wang, L., Xu, J.: Using post-harvest waste to improve shearing behaviour of loess and its validation by multiscale direct shear tests. Appl. Sci. **9**(23), 5206 (2019)

Duan, Z., Cheng, W.-C., Peng, J.-B., Wang, Q.-Y., Chen, W.: Investigation into the triggering mechanism of loess landslides in the south Jingyang platform, Shaanxi province. Bull. Eng. Geol. Env. **78**(7), 4919–4930 (2018). https://doi.org/10.1007/s10064-018-01432-8

Dinh, N.H., Tran, H.V., Choi, K.K.: Direct shear behavior of cementitious mortar reinforced by carbon fiber textile. Construc. Build. Mater. **249**, 118760 (2020)

Fan, X.M., Xu, Q., Scaringi, G., Li, S., Peng, D.L.: A chemo-mechanical insight into the failure mechanism of frequently occurred landslides in the Loess Plateau, Gansu Province, China. Eng. Geol. **228**, 337–345 (2017)

Fatahi, B., Khabbaz, H., Fatahi, B.: Mechanical characteristics of soft clay treated with fibre and cement. Geosyn. Int. **19**, 252–262 (2012)

Gray, D.H., Barker, D.: Root-soil mechanics and interactions. Water Sci. Appl. **8**, 113–123 (2004)

Ghazizadeh, S., Bareither, C.A.: Stress-controlled direct shear testing of geosynthetic clay liners I: apparatus development. Geotext. Geomemb. **46**, 656–666 (2018)

Gu, T.F., Zhang, M.S., Wang, J.D., Wang, C.X., Xu, Y.J., Wang, X.: The effect of irrigation on slope stability in the Heifangtai Platform, Gansu Province, China. Eng. Geol. **248**, 346–356 (2019)

Guo, W.Z., et al.: Telling a different story: the promote role of vegetation in the initiation of shallow landslides during rainfall on the Chinese Loess Plateau. Geomorphology **350**, 106879 (2020)

Hungr, O.: A model for the runout analysis of rapid flow slides, debris flows and avalanches. Can. Geotech. J. **32**, 610–623 (1995)

Hungr, O.: Rock avalanche occurrence, process and modelling. Landslides **302**, 243–266 (2006)

Kua, T., Arulrajah, A., Mohammadinia, A., Horpibulsuk, S., Mirzababaei, M.: Stiffness and deformation properties of spent coffee grounds based geopolymers. Constr. Build. Mater. **138**, 79–87 (2017)

Li, Y.R., Mo, P.: A unified landslide classification system for loess slopes: a critical review. Geomorphology **340**, 67–83 (2019)

Li, P., Shen, W., Hou, X.K., Li, T.L.: Numerical simulation of the propagation process of a rapid flow-like landslide considering bed entrainment: a case study. Eng. Geol. **263**, 105287 (2019)

Luo, H., Wu, F.Q., Chang, J.Y., Xu, J.B.: Microstructural constraints on geotechnical properties of Malan Loess: a case study from Zhaojiaan landslide in Shaanxi province, China. Eng. Geol. **236**, 60–69 (2018)

Leng, Y.Q., Peng, J.B., Wang, Q.Y., Meng, Z.J., Huang, W.L.: A fluidized landslide occurred in the Loess Plateau: a study on loess landslide in South Jingyang tableland. Eng. Geol. **236**, 129–136 (2018)

Lian, B., Peng, J., Zhan, H., Wang, X.: Mechanical response of root-reinforced loess with various water contens. Soil Tillage Res. **193**, 85–94 (2019)

Modoni, G., Croce, P., Mongiovi, L.: Theoretical modelling of jet grouting. Geotechnique **56**(5), 335–347 (2006)

Modoni, G., Bzòwka, J.: Analysis of foundation reinforced with jet grouting. J. Geotech. Geoenviron. Eng. **138**(12), 1442–1454 (2012)

Modoni, G., Flora, A., Lirer, S., Ochmański, M., Croce, P.: Design of jet grouted excavation bottom plugs. J. Geotech. Geoenviron. Eng. **142**(7), 04016018 (2016)

Mangeney, A., Roche, O., Hungr, O., Mangold, N., Faccanoni, G., Lucas, A.: Erosion and mobility in granular collapse over sloping beds. J. Geophys. Res. Earth Surface **115**, 3–40 (2010)

Nguyen, L., Fatahi, B.: Behaviour of clay treated with cement & fibre while capturing cementation degradation and fibre failure – C3F Model. Int. J. Plast **81**, 168–195 (2016)

Pei, X.J., Zhang, X.C., Guo, B., Wang, G.H., Zhang, F.Y.: Experimental case study of seismically induced loess liquefaction and landslide. Eng. Geol. **223**, 23–30 (2017)

Peng, J.B., et al.: Heavy rainfall triggered loess-mudstone landslide and subsequent debris flow in Tianshui, China. Eng. Geol. **186**, 79–90 (2015)

Peng, D.L., et al.: Hydrological response of loess slopes with reference to widespread landslide events in the Heifangtai terrace, NW China. J. Asian Earth Sci. **171**, 259–276 (2019)

Peng, J.B., et al.: Landslides triggered by excavation in the loess plateau of China: a case study of Middle Pleistocene loess slopes. J. Asian Earth Sci. **171**, 246–258 (2019)

Peng, J.B., et al.: Interaction between landsliding materials and the underlying erodible bed in a loess flowslide. Eng. Geol. **234**, 38–49 (2018)

Peng, J., Tong, X., Wang, S., Ma, P.: Three-dimensional geological structures and sliding factors and modes of loess landslides. Environ. Earth Sci. **77**(19), 1–14 (2018). https://doi.org/10.1007/s12665-018-7863-y

Peerun, M.I., Ong, D.E.L., Choo, C.S.: Interpretation of geomaterial behaviour during shearing aided by PIV technology. J. Mater. Civ. Eng. **31**(9), 04019195 (2019)

Shen, S.L., Wang, Z.F., Horpibulsuk, S., Kim, Y.H.: Jet-grouting with a newly developed technology: twin-jet method. Eng. Geol. **152**(1), 87–95 (2013)

Shen, S.L., Wang, Z.F., Sun, W.J., Wang, L.B., Horpibulsuk, S.: A field trial of horizontal jet grouting with composite-pipe method in soft deposit of Shanghai. Tunn. Undergr. Space Technol. **35**, 142–151 (2013)

Shen, S.L., Wang, Z.F., Yang, J., Ho, E.C.: Generalized approach for prediction of jet grout column diameter. J. Geotech. Geoenviron. Eng. **139**(12), 2060–2069 (2013)

Sivakumar Babu, G.L., Raja Jaladurgam, M.E.: Strength and deformation characteristics of fly ash mixed with randomly distributed plastic waste. J. Mater. Civ. Eng. **26**(12), 04014093 (2014)

Shi, J.S., Wu, L.Z., Wu, S.R., Li, B., Wang, T., Xin, P.: Analysis of the causes of large-scale loess landslides in Baoji, China. Geomorphology **264**, 109–117 (2016)

Suksiripattanapong, C., Kua, T., Arulrajah, A., Maghool, F., Horpibulsuk, S.: Strength and microstructure properties of spent coffee grounds stabilized with rice husk ash and slag geopolymers. Constr. Build. Mater. **146**, 312–320 (2017)

Shen, S.L., Wang, Z.F., Cheng, W.C.: Estimation of lateral displacement induced by jet grouting in clayey soils. Geotechnique **67**(7), 621–630 (2017)

Tian, J., Huang, Y., Yong, S.: The soil physical properties with relation to the loess genesis in the loess region. Memoir NISWC Academia Sinica **5**, 1–12 (1987)

Tabarsa, A., Latifi, N., Meehan, C.L., Manahiloh, K.N.: Laboratory investigation and field evaluation of loess improvement using nanoclay – a sustainable material for construction. Constr. Build. Mater. **158**, 454–463 (2018)

Tan, H., Chen, F., Gao, Y.: Direct shear tests of shear strength of soils reinforced by geomats and plant roots. Geotext. Geomembr. **47**, 780–791 (2019)

Wang, X.G., Wang, J.D., Zhan, H.B., Li, P., Qiu, H.J., Hu, S.: Moisture content effect on the creep behavior of loess for the catastrophic Baqiao landslide. CATENA **187**, 104371 (2020)
Wang, G.H., Zhang, D.X., Furuya, G., Yang, J.: Pore-pressure generation and fluidization in a loess landslide triggered by the 1920 Haiyuan earthquake, China: a case study. Eng. Geol. **174**, 36–45 (2014)
Wang, S.K., Peng, J.B., Zhuang, J.Q., Kang, C.Y., Jia, Z.J.: Underlying mechanisms of the geohazards of macro Loess discontinuities on the Chinese Loess Plateau. Eng. Geol. **263**, 105357 (2019a)
Wang, H.L., Zhou, W.H., Yin, Z.Y., Jie, X.X.: Effect of grain size distribution of sandy soil on shearing behaviors at soil-structure interface. J. Mater. Civ. Eng. **31**(10), 04019238 (2019)
Xu, L., Dai, F.C., Tham, L.G., Zhou, Y.F., Wu, C.X.: Investigating landslide-related cracks along the edge of two loess platforms in northwest China. Earth Surf. Proc. Land. **37**(10), 1023–1033 (2012)
Xu, L., et al.: Occurrence of landsliding on slopes where flowsliding had previously occurred: an investigation in a loess platform, North-west China. CATENA **104**, 195–209 (2013)
Xu, L., Coop, M.R., Zhang, M.S., Wang, G.L.: The mechanics of a saturated silty loess and implications for landslides. Eng. Geol. **236**, 29–42 (2018)
Xu, R., Li, X., Yang, W., Jiang, C., Rabiei, M.: Use of local plants for ecological restoration and slope stability: a possible application in Yan'an, Loess Plateau, China. Geomatics Natural Hazards Risk **10**(1), 2106–2128 (2019)
Yoobanpot, N., Jamsawang, P., Horpibulsuk, S.: Strength behavior and microstructural characteristics of soft clay stabilized with cement kiln dust and fly ash residue. Appl. Clay Sci. **141**, 146–156 (2017)
Zaimoglu, S.: Freezing–thawing behavior of fine-grained soils reinforced with polypropylene fibers. Cold Regions Sci. Tech. **60**(1), 63–65 (2010)
Zaimoglu, A.S., Yetimoglu, T.: Strength behavior of fine grained soil reinforced with randomly distributed polypropylene fibers. Geotech. Geol. Eng. **30**(1), 197–203 (2012)
Zhang, F.Y., Wang, G.H.: Effect of irrigation-induced densification on the post-failure behavior of loess flowslides occurring on the Heifangtai area, Gansu, China. Eng. Geol. **236**, 111–118 (2017)
Zhuang, J., et al.: Prediction of rainfall-induced shallow landslides in the Loess Plateau, Yan'an, China, using the TRIGRS model. Earth Surf. Proc. Land. **42**, 915–927 (2017)
Zhang, S., Pei, X.J., Wang, S.Y., Huang, R.Q., Zhang, X.C., Chang, Z.L.: Centrifuge model testing of a loess landslide induced by rising groundwater in Northwest China. Eng. Geol. **259**, 105170 (2019)
Zhang, C., Chen, L., Liu, Y., Ji, X., Liu, X.: Triaxial compression test of soil-root composites to evaluate influence of roots on soil shear strength. Ecol. Eng. **36**, 19–26 (2010)

Enhancement of Expansive Soils with the Treatment of Lignosulfonate

Richa Mudliar(✉) and Rajesh Sathiyamoorthy

Department of Civil Engineering, IIT Kanpur, Kanpur 208016, India
{vstricha,hsrajesh}@iitk.ac.in

Abstract. Expansive soils possess swelling behavior on the absorption of water, causing heaving on the ground surface. Two artificial highly-plastic soils are tested for the enhancement in engineering properties after treatment with a hydrophobic chemical. The water-proofing nature of the chemical resulted in the reduction of swelling and shrinking behavior. This study involves the evaluation of swell pressure, compressibility, strength on the dry and wet side of optimum moisture content. Shrinkage behavior is evaluated based on linear shrinkage test done at 2% higher than the liquid limit of the soil. The optimum dosage of lignosulfonate is decided based on the swell-shrinkage characteristics with different variation of the chemical. Crack intensity factor obtained from linear shrinkage tests is decreased by 50% after the addition of optimum chemical content. For soil A with a liquid limit of 140.94%, the optimum chemical mix is at 3% of the mass of soil while for soil B with a liquid limit of 105%, it is at 2% chemical.

1 Introduction

The expansive soil undergoes swelling and shrinkage due to volumetric instability when exposed to moisture. The volumetric instability leads to excessive settlements or heaving due to uneven soil movement causing damage to structures. Expansive soil mainly consists of reactive montmorillonite mineral having open sites for the attraction of positive charge which is neutralized by the absorption of mono- or divalent cations. Bolt (1956) concluded that swelling of minerals present in clays is due to excess osmotic pressure in the adsorbed layer of ions. It is the only internal pressure acting between particles. When the soil is subjected to external pressure, the distance between particles will decrease and water will be squeezed out. As a result, the ion concentration between the particles will increase and the osmotic pressure in turn increases. Equilibrium is finally reached when the osmotic pressure equals the external pressure. The reverse process involves the decrease of vertical stress and the suction of water by osmotic pressure between the particles to dilute the concentration of ions. The distance between the particles increases, resulting in volume increase and a reduction of osmotic pressure. The process continues until a new equilibrium is established. The imbibitions of water are the most important cause of swelling. Moisture heave sometimes takes place in heavily consolidated clays with high plasticity index. In many cases this is a matter of the reswelling of montmorillonite which has been reduced to relatively low moisture content by consolidation stress. The main problems to be considered while dealing with expansive soil

B. Fatahi et al. (Eds.): GeoChina 2021, SUCI, pp. 163–177, 2021.
https://doi.org/10.1007/978-3-030-79854-3_12

are swelling and permeability characteristics. Since expansive soils pose the damage through cyclic swelling and shrinkage in the field leading to fatigue of the strength of the structure, it is important to consider before designing the structure. Permeability has been one of the major issues in the field, causes the problem of ponding water due to the low permeability of the soil. For achieving the desirable geotechnical properties, stabilization is the economical and long-lasting method. Soil stabilization through traditional stabilizers such as lime, cement, fly-ash, have become main concerns for damage to the environment, via the release of greenhouse gases or producing leachate of chemicals in groundwater (Jayanthi and Singh 2016).

However, non-traditional chemicals including polymers, resins, enzymes, salts have become the spotlight of research (Tingle and Saltoni 2003). Non-traditional stabilizers like lignosulfonate have been reported to improve flocculation or aggregation structure from the repulsive dispersion nature of clay particles (Alazigha et al. 2017). Lignosulfonate (LS) is known as a sustainable by-product from the pulp industry and therefore is an economical and widely available chemical. The optimum dosage of lignosulfonate may need to be decided based on tests like swell, permeability, consolidation, unconfined compressive strength, and CBR tests. Negonbadi and Noorzad (2017) performed wetting-drying cycles on highly plastic clays treated with lignosulfonate for direct shear and unconsolidated undrained triaxial tests. It was reported that with the increase in wet-dry cycles, shear strength parameters (cohesion and angle of internal friction) do not change significantly but at the end of the 4th cycles, the cohesion has increased by 30 kPa proving the efficiency of chemical in increasing strength.

It is quite clear, that lignosulfonate has proved to have clay-amendment properties. However, literature related to the effect of LS on the treatment of expansive soils is limited and are not consistent in defining the stabilization scheme. In particular, properties like shrink behavior, crack properties, and unconfined compressive strength under different conditions need to be explored. Moreover, the efficiency of the chemical depends on the clay content present in the soil. Hence, there is a need to establish a beneficial aspect of adding LS to the expansive soil. The objective of the present study is to investigate the efficiency of lignosulfonate for the stabilization of expansive soils. Various engineering properties of expansive soils with different dosages of LS were evaluated. From the test results, the optimum dosage of the chemical needed for enhancing the strength and deformation of expansive soil was quantified.

2 Materials Used and Methodology

2.1 Materials Used

Artificial soil with bentonite and fly-ash is selected based on the basic properties and swell (%). Previously researchers have also studied the swelling behavior of bentonite (Villar and Lloret 2008), fly ash (Brandl 1992), bottom ash–bentonite mixtures (Pal and Ghosh 2014), and fly ash–stabilized expansive clay (Phani Kumar and Sharma 2007). For the present study, bentonite -fly ash mix proportion of 1:4 by dry weight (termed as Soil A) and 1:3 (termed as Soil B) is chosen for testing the efficacy of lignosulfonate for expansive soils. The mix proportion is decided based on the swelling and permeability property of the mix (Younus and Sreedeep 2012, Mollamahmutoğlu and Yilmaz 2001).

Bentonite is very fine clay; more than 60% of particles have a diameter of less than 2 mm. Liquid limit (LL) and plastic limit (PL) of the bentonite are 574% and 66%,respectively. The PI of bentonite indicates that the bentonite is a very highly plastic material and has high swelling potential. Fly-ash used in the study has been collected from Panki thermal power plant and has 8% clay fraction. After mixing in respective proportions the bentonite and fly ash, the index and engineering properties are summarized in Table 1 according to ASTM standards.

The calcium composition quantitatively exceeds other elements present. This implies that calcium is the dominant salt present in the LS, thus the LS is a calcium lignosulfonate (CaLS). The chosen dosage in lignosulfonate is based on index properties 0, 1, 3, 5% LS for soil A and 0, 1, 2, 3% LS for soil B. The calculated amount of Ca-lignosulfonate was mixed with water to prepare the LS solution of the chosen dosage.

Table 1. Index and Engineering properties of expansive soil

Property	ASTM code used for evaluation	Bentonite	Fly-ash	Soil A (20B80FA)	Soil B (25B75FA)
Sand content	D6913-04		56%	39%	35%
Silt content	D6913-04	16%	36%	46%	43%
Clay content	D6913-04	84%	8%	17%	22%
Liquid limit (%)	IS-2720-part-5-1985	574%		108	140.94
Plastic limit (%)	D4318-17	66%		40	47.12
Shrinkage limit (%)	D4943-18			20	59.65
Maximum dry density (gm/cc)	D698-12			1.2	1.17
Optimum moisture content (%)				29.22	31.44
Swell (%)	D4546-14			9.9% (OMC–2%); 2.8% (OMC + 2%)	15% (OMC–2%); 5% (OMC + 2%)
Compression index	D2345-11			0.473 (OMC–2%); 0.172 (OMC + 2%)	0.76 (OMC–2%); 0.52 (OMC + 2%)
Unconfined Compressive Strength (kPa)	D2166-13			416.8 (OMC–2%) 315.28 (OMC + 2%)	293 kPa (OMC–2%) 179.3 kPa (OMC + 2%)

2.2 Experimental Procedure

Compaction tests carried out according to the ASTM D698-12e2 standard on both the soils using standard Proctor compaction energy. The required amount of LS is added to the water content corresponding to the state of compaction for mechanical tests.

The atterberg limits for the treated sample is found after one hour of equilibrium kept after mixing with the lignosulfonate. For this, the required amount of chemical is mixed with amount of water to get consistency in soil. The solution is then added to the soil, mixed and then kept in equilibrium for an hour followed by cone penetration test. The sequence of mixing here will not impact the results as trial experiments on atterberg limits have given same values. The sequence of mixing is fixed based on the fact that lignosulfonate is soluble in water (Soltani et al. 2019; Alazigha et al. 2018). And doesn't release any gases such as carbon dioxide on reaction with water contrary to the case of lime.

The soil and the prepared LS solution were mixed thoroughly and compacted. Various tests conducted are atterberg limit tests, compaction tests, one dimensional swell-consolidation test, linear shrinkage test, unconfined compressive test.

2.2.1 1-D Swell Consolidation Test

One-dimensional free swell tests were carried out on natural and stabilized soil samples according to the ASTM D4546-14 standard, and under a constant surcharge pressure of 5 kPa. Samples (20 mm diameter and 60 mm height) were placed in the oedometer and inundated with DI water. Compaction is carried out in two layers by dynamic compaction. Axial deformation or swelling was then recorded during various stages of swelling to a point in which axial swelling attained an equilibrium value defined as swelling potential or free swell percent. For a given elapsed time of swelling, axial swelling was determined by the following Eq. (1):

$$(\%)\varepsilon_a(t) = \frac{\Delta H(t)}{H_0} * 100 \tag{1}$$

where $(\%) \in_a (t)$ is axial swelling in respect to elapsed time t (%), $\Delta H(t)$ is increase in the height of the soil sample in respect to elapsed time t, and H(t) is initial height of the soil sample or height of the oedometer ring (20 mm).

Once the dial deformation attains the equilibrium value, vertical pressure is loaded in increments of 10 kPa, 20 kPa, 50 kPa, 100 kPa, 200 kPa, 400 kPa, 800 kPa. The standard load increment duration should exceed the minimum completion of primary consolidation. The maximum vertical stress must be sufficient enough to get three points in a straight line for e-logp curve.

2.2.2 Unconfined Compressive Test

Cylindrical soil samples of 38 mm in diameter and 76 mm in height are used for unconfined compressive strength test. For treated samples, the required amount of chemical is mixed with water according to dry side (OMC−2%) and wet side (OMC + 2%). The solution is added to the soil taken according to dry-density corresponding to the water content. Sample is prepared in five layers by dynamic compaction. Samples are kept for curing for 28 days at temperature of 20 $\pm$ 5C. while the untreated samples are kept Shearing is done after curing of samples, until failure by deforming the sample at the rate of 0.210 mm/min for determination of UCS.

2.2.3 Linear Shrinkage Test

Linear shrinkage is defined as change in length of the specimen, expressed in percentage of its initial length when sample is allowed to dry from specific water content to shrinkage limit, Eq. (2). Linear shrinkage test was performed as per BS 1377–2 (1990). A fabricated semi-cylindrical mould made up of brass having length 140 mm and diameter 25 mm used for this test. The inner surface of the mould used was smooth to ensure that the friction between the soil layer and mould becomes negligible. Soil is mixed with water (LL + 2%) to make a consistent paste and is filled in the mould. Care has been taken no air bubbles remain in the sample. Sample is subjected to air-drying for 24 h followed by oven-drying till the weight becomes constant. After drying of specimen, the final images are taken. The images are converted to grey scale image from RGB. Thresholding is done converting the image into binary form. Linear shrinkage test is conducted on untreated and stabilized soils to observe the effect with variation in dosage of chemical. Crack intensity factor (CIF) is the quantitative measurement for cracked area (black portion in the image) with respect to the total area of the specimen, Eq. (3) For the purpose of calculation of CIF, the sample is allowed to dry in air and images are taken at initial and final stage of drying. Parameters from the test are.

$$\text{Linear shrinkage, } LS(\%) = \frac{L_s}{L} * 100 \tag{2}$$

$$\text{Crack intensity factor, } CIF(\%) = \frac{A_C}{A} * 100 \tag{3}$$

3 Results and Discussion

3.1 Effect of Lignosulfonate on Plasticity

The liquid limit is defined as the water content corresponding to the state of soil yielding shear strength of 1.7 kPa, Wroth and Wood (1985). Figure 1(a) shows the variation of the liquid limit at various dosages of LS chemical for soil A and soil B. With increase in LS content, liquid limit decreases upto definite amount of LS and then becomes constant. Table 1 shows that Soil B has 5% higher clay content than soil A, consequently effect of plasticity can be seen for the increase in liquid-limit of the soil B. A drop in the liquid limit is observed for the dosage of 1% LS for both the soils. For Soil A, liquid limit decreases drastically on addition of 1% LS from 110% to 60%, rate of reduction decreases with further addition of LS. The plastic limit of soil at different dosages of the chemical was found to be almost constant, hence the plasticity index decreases with the increase in the chemical dosage. For soil B, reduction of 50% is observed at 1% addition of LS. On increasing the chemical dosage from 1% to 5%, further reduction is observed from 100% to 53.5%. Beyond 5% LS, no change in liquid limit is observed. Soil A has little dip in liquid limit at 3%LS and then there is increase of 10% LL at 4% LS with no change on further addition of LS. This can be visualized as LS behaves like an expanding polyelectrolyte. Positive charge in the clay gets neutralized by the strong negative charges present in the LS. So, near the optimum content of LS, the excess molecules of LS are thrown out as the result of repulsive forces which causes expansion (Alazigha et al. 2018). While in case of soil B, no dip is observed in LL with variation in chemical content.

Figure 2 shows the plasticity chart with the position of the treated soils. Due to decrease in liquid limit with clay content, for soil A the soil classification changed from CH (with LL = 108%, PI = 68%) to MH (2% LS) and MI-MH (3% LS) while the decrease in plasticity index of soil B lead to change in soil type from CH to MH (1% LS) with no change in soil type with increase in the chemical dosage, reflecting the significance of LS in reducing the plasticity nature of the soil. The change in the plasticity index reflected as change in soil type due to transformation of clay particle into silty with change in particle size.

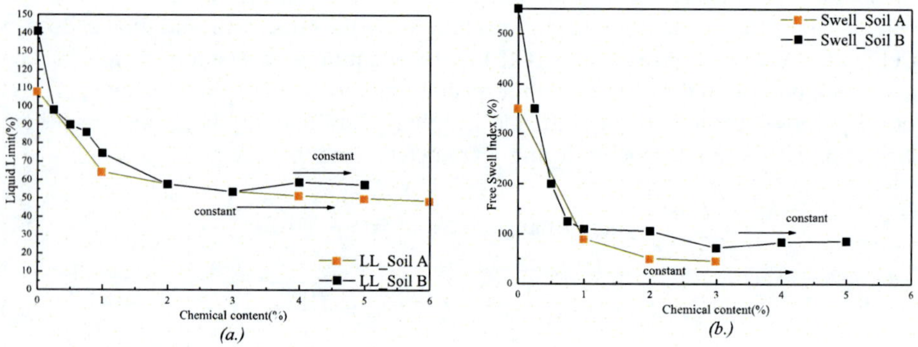

Fig. 1. Variation of (a) liquid-limit and (b) free swell index with chemical content

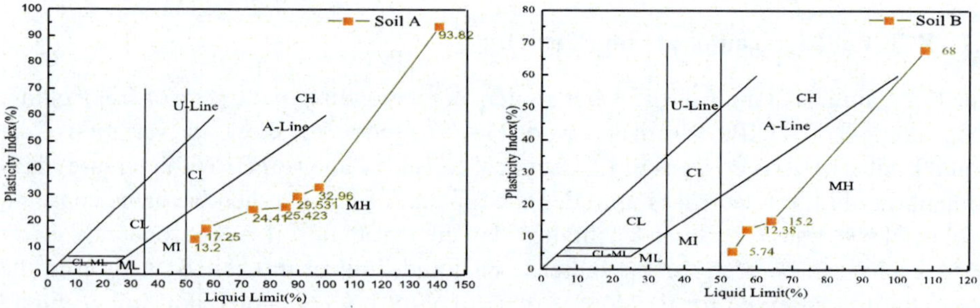

Fig. 2. Soil classification in plasticity chart at different dosages of LS

3.2 Effect of Lignosulfonate on Swelling Property of Soil

The variation of the FSI of soil with different dosages of LS is shown in Fig. 1(b) for both the soils. The value of FSI directly proportional to the amount of montmorillonite present in the soil. For the soil blended with fly-ash, the presence of fly-ash changes the clay particle sizes through pozzolanic reactions. The reaction of lignosulfonate with fly-ash increases the amount of calcium carbonate produced. For soil A, the change in particle size of clay particles carried out till 5% LS beyond which, repulsive behavior of coated particles lead to increase in FSI. It can be noticed that FSI follows the same trend as the liquid limit, which implies, swelling nature of the expansive soil can be controlled with the addition of LS but the extent of reduction in swelling depends on the dosage of chemicals and clay content.

Soils are compacted at both sides of optimum water content of compaction curve to investigate the usage at different fields according to requirements. Study of swelling on two different swelling soils has been conducted with surcharge of 5 kPa. Bolt (1956) suggests that osmotic pressure indeed develops in the soil—water system is responsible for the swelling mechanism. It can be observed the dry side of optimum has more swell (%) than wet side of optimum. This is due to the fact that dry side of optimum has flocculated structure while wet side of optimum has dispersed structure. Irrespective of the amount of water, the swell (%) of soils decreased with increase in LS content. The decrease in swell (%) is due to decrease in diffuse double layer thickness. The presence of di-cations in Ca-LS react with the open sites of clay minerals negative charge. According to Holtz and Gibbs, 1956, the soil under surcharge of 7 kPa posing the swell (%) of more than 10% can be categorized into medium swell, while greater 20% into high swelling soils. For soil A, the swell (%) was 10% at dry side of optimum and 3% at wet side, Fig. 3(a). The dry side of optimum can be categorized into medium swell soil, while wet side in low swell soil. The swelling (%) found to decrease upto 3% from 10%, after 3% LS swell (%) attains constant value. At 2% of the chemical, a significant reduction in the percent swell can be noticed at the dry side of optimum for soil A. At wet-side of optimum, percent swell is observed to be negligible at 2% and 3% LS. The soil B (Fig. 3(b)) having high percent swell compared to soil B due to high clay-content. The percent swell attains a constant value beyond 5% LS for soil for soil B, similar to liquid

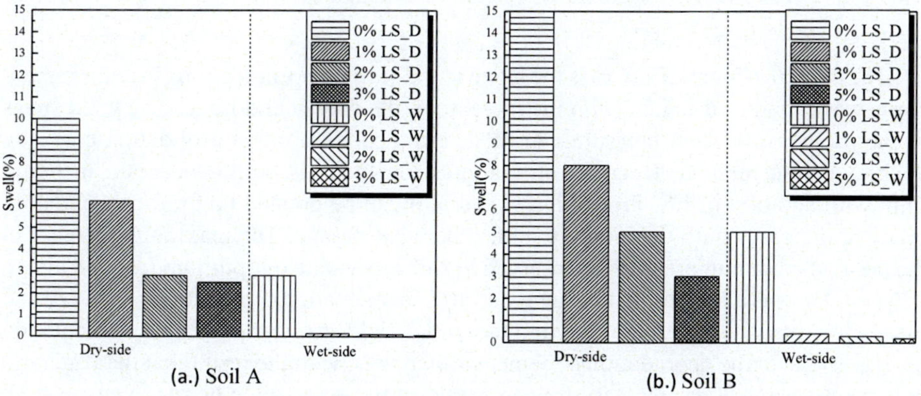

Fig. 3. Swell (%) of soils with different dosages of LS

limit. There is dip in swell (%) in 3% LS in dry side of optimum similar to FSI and LL variation with chemical content.

3.3 Effect of LS on Consolidation Characteristics of Soils

The swell-consolidation behavior of soil A and soil B compacted at wet and dry side of optimum is shown in Figs 4 and 5. The swell pressure (Ps) is the vertical pressure required to recompress the void ratio of the soil at swollen state to the initial state of compaction. Table 2 provides a summary of the swell pressure and the compression index of the untreated and treated soil at various placement conditions. From Table 2, the swelling pressure of soil B is higher than soil A due to higher amount of bentonite in soil B than A. The swelling pressure of untreated soil is higher at the dry side of optimum as compared to the wet side of optimum. With the addition of LS, the swell pressure of soil has been drastically reduced for both the soils. . For soil A, the maximum reduction in the swell pressure at the dry side of optimum 91.92% was noticed for 3% of LS while at the wet side of optimum maximum reduction was noticed at 3% of LS as compared to the untreated soil. The reduction in swell pressure for soil B is 50% at 1% LS addition, and reduced with subsequent increase in amount of chemical by 75% (5% LS) in dry side of optimum.

Table 2. Consolidation characteristics of soils

Chemical dosage (%)	Soil A				Soil B			
	Dry side		Wet side		Dry side		Wet side	
	Ps	Cc	Ps	Cc	Ps	Cc	Ps	Cc
0	161	0.473	30	0.172	200	0.76	110	0.52
1	30	0.127	14	0.16	102	0.42	70	0.34
2	13.6	0.116	10	0.129	90	0.28	50	0.25
3	10	0.104	10	0.084	-	-	-	-
5	-	-	-	-	40.5	0.22	30	0.18

-Not performed, Ps-swell pressure (kPa), Cc-compression index

The compression index (Cc) is the slope of the straight portion of virgin compression curve was found to decrease with the increase in the chemical dosage. The yield stress or the pre-consolidation pressure can be determined from the consolidation curve by double-tangent method. Decrease in compression index is due to decrease in liquid limit with addition of LS. From Figs. 4 and 5, it can be noticed that pre-consolidation pressure increases with the increase in the chemical dosage. The maximum increase in the pre-consolidation pressure is observed in soil A dry-side of optimum from 48 kPa to 176 kPa. There is a possibility that at higher stress cementation degradation occurs. As the mean effective stress increases beyond the initial yield stress, the effect of cementation decreases due to the degradation of cementation bonds, with the result that the void ratio of cement-treated clays decreases due to breakage of cementation bonds (Nguyen et al.

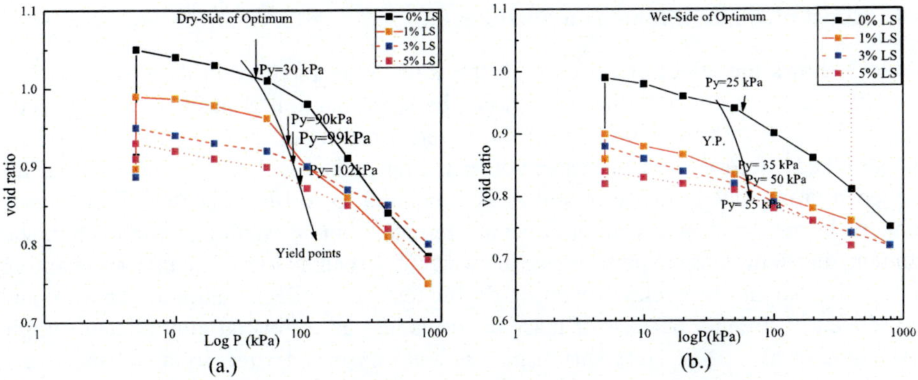

Fig. 4. Consolidation characteristics of soil A with varying dosage of LS

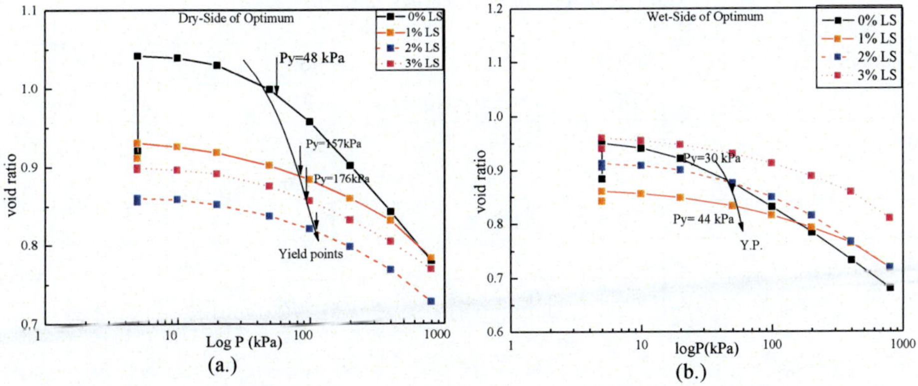

Fig. 5. Consolidation characteristics of soil B with different dosage of LS

2017). The maximum increase observed in soil A is expected, as the soil contains 80% silt content in the form of fly-ash. Lignosulfonate forms coating layer over remaining particles of clay resulting in increase in diameter of particles and consequently change the soil type from CH to MH. At the wet and dry side of optimum was found to be 25% and 72.72% at LS percentage of 3% in soil A. While in case of soil A, increase of 70 kPa is seen at 5% LS at dry-side of optimum and 20 kPa at wet side of optimum.

The possible mechanism for stabilization is cations replacement (Alazigha et al. 2016). Lignosulfonate undergoes cations exchange and peripheral adsorption on the surfaces of the montmorillonite minerals. This results in smearing and subsequent aggregation of soil particles. The reduction in the number and volume of intra-aggregate pore spaces compensated by the formation of larger but fewer inter-aggregate pores leads to reduction in permeability.

3.4 Effect of Lignosulfonate on Shrinkage

Table 3 shows the effect of LS on both the soils. With increase in the dosage of LS, the CIF decreases for both the soils. It can be seen that soil B has more pronounced results as compared to soil A. Since the liquid limit of soil B is higher than soil A, the shrinkage observed for untreated soil B is higher than soil A. The influence of LS has been pronounced in case of soil A as compared to soil B. It is clear from Fig. 6, that cracks are decreasing with increase in chemical content. With increase in chemical content, due to increase in particle size the water absorbed by the soil decreases resulting in less shrinkage. The cycles of wetting-drying test lead to degradation of chemical and subsequently strength but loss of mass of soil is less as compared to untreated sample (Alazigha et al. 2017). Here shrinkage has been observed after drying of the slurry sample.

Table 3. Effect of LS on crack intensity factor

Chemical content (%)	Soil A		Soil B	
	CIF	Linear shrinkage	CIF	Linear shrinkage
0	17.6	10.14	25.458	13.57
1	6.4	3.2	21.6	3.5
2	3.1	2.1	-	
3	2.4	0.1	11.34	2.2
5	-	0	8.65	0.9

It can be observed that with increase chemical content, the tendency of water loss decreases due to decrease in absorption of water. The particle size increases and therefore, the DDL decreases resulting in increase in volume of solids as compared to untreated sample. Therefore, the clay content decreases with increase in particle size.

Soil A has lower liquid limit as compared to soil B. The water absorption capacity of soil B is more resulting in larger water phase as compared to soil A. Therefore, the crack intensity factor as well as linear shrinkage due to higher suction stress developed during shrinkage process. It can be clearly seen from Fig. 6(a) and (b), the black area (cracked area) in case of soil B is more than soil A. The crack pattern developed in between may have same area, but the width of crack in case of soil B is more than soil A. The linear shrinkage can be measured as length of sample after drying. The width of the crack in case of Fig. 6(c) and (d) can be clearly observed to conclude that with increase in chemical content, the black area occupied by shrinkage part decreases significantly. In case of 3% LS, the difference between soil B and soil A is still significant of about 9%. It can be concluded that soil B CIF doesn't get reduced after a certain value and is more than soil A, due to higher clay content present.

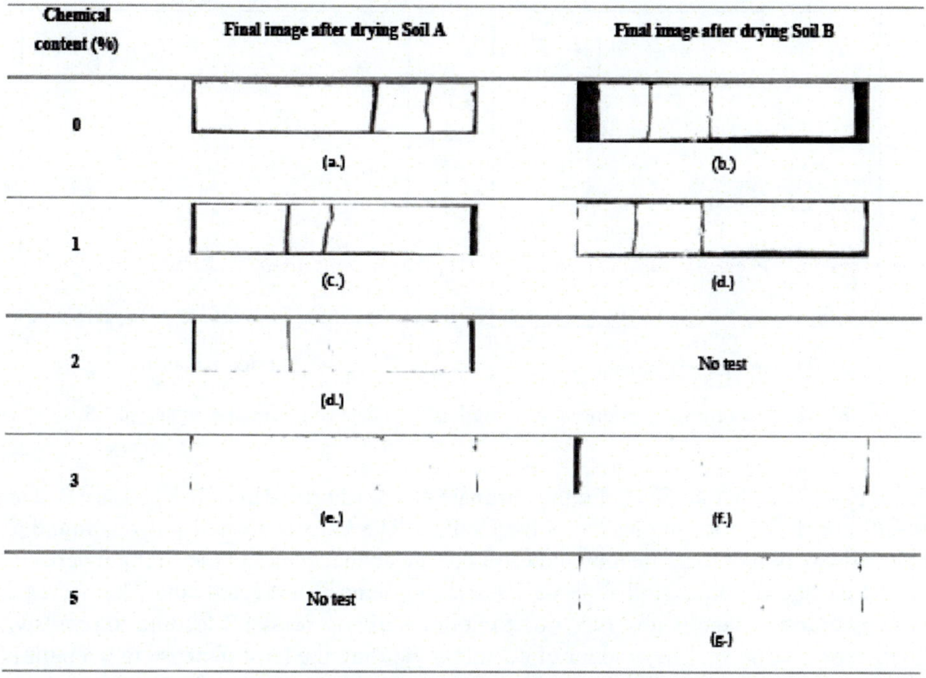

Fig. 6. Processed image after final drying

3.5 Effect of LS on Unconfined Compressive Strength of Soils

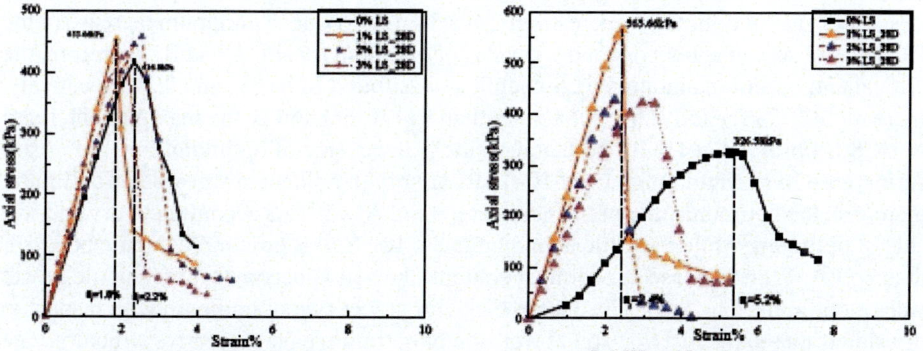

(a.) Dry side of optimum

(b.) Wet side of optimum

Fig. 7. Unconfined compressive strength of soil A at different percentage of LS

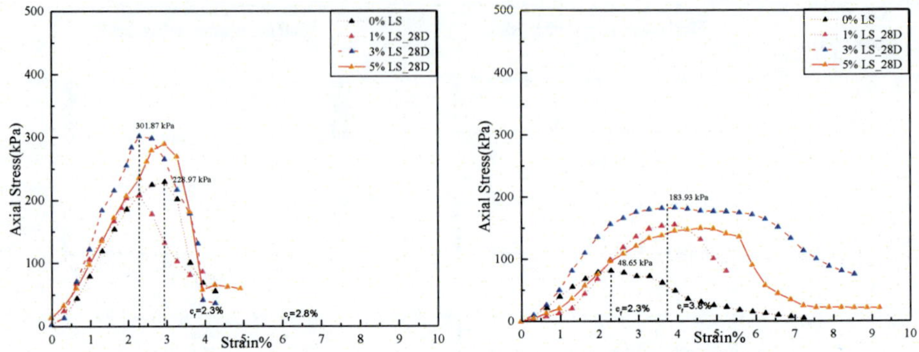

(a.) Dry side of optimum (b.) Wet side of optimum

Fig. 8. Unconfined compressive strength of soil B at different percentage of LS

From Figs. 7 and 8, the effect of lignosulfonate can be observed on soil A and soil B. The results are shown for samples cured for 28 days. The UCS of treated soil was found to be relatively higher than the untreated soil but the enhancement in the strength depends on the dosage of the chemical as well as clay content of untreated soil. The decrease after a certain dosage is observed due to non-absorbed excess LS leading to repulsive behavior of particles. The amount of chemical yielding the least increase in strength is termed as the critical amount of chemical, while the amount of chemical having highest rise in strength is termed as the optimum amount of chemical.

For soil B, UCS for untreated sample is 100 kPa less than the soil A. For soil A, at the LS dosage of 1% and 2%, the maximum increase in the UCS of soil compacted at the dry and the wet side of optimum was found to be 50 kPa and 200 kPa respectively, Fig. 7. Similarly, at a 2% LS dosage gives increase of 40 kPa at dry side of optimum and 100 kPa at wet side of optimum. Interestingly, at a 3% LS dosage, a minimum increase in the UCS was observed at both dry and wet sides of optimum for soil A and B. Therefore, the critical and optimum amount of LS chemical was found to be 3% and 2% respectively. Increase in UCS for soil B has been more than soil A. For soil B, the maximum increase in UCS is observed at 1% LS both at dry side and wet side of optimum, Fig. 8. For soil A, increase in strength range from 10-0 kPa at both dry side and wet side of optimum. therefore, the optimum amount of chemical for soil A is 3% LS according to dry and wet side of optimum, while the critical amount is 5% LS. Soil B pose more ductile behavior than soil A. With increase in chemical content, the soil B becomes more ductile at wet side of optimum, Fig. 8(b). The ductile behavior at wet side of optimum with respect to dry side is due to the fact that soil at wet side of optimum have dispersive structure. The failure strain for the wet side of optimum is at higher side as compared to dry side of optimum.

3.6 Optimum Dosage of LS

The optimum dosage of lignosulfonate for effective stabilization of expansive soils can be decided based on engineering properties swell (%), compression index (Cc), strength (UCS) and crack intenity factor (CIF). The optimum dosage of LS in terms of percentage swell can be found from Fig. 3. From Fig. 3, swell (%) decreases for soil B drops at 1% LS for both dry and wet-side of optimum. The maximum decrease of swell (%) upto 3% for 5% LS is observed. While for wet-side of optimum, swell (%) becomes negligible on addition of 3% and 5% LS. The compression index is observed to decrease with increase in dosage of chemical. For soil A, the percentage swell is found to be constant beyond 2% LS for both dry and wet side of optimums. Interestingly, with the addition of LS to the swelling soil, the compression index of treated soil was found to be less than the permissible value, irrespective of the dosage of chemical and placement conditions (Table 2). A minimum dosage of 1% LS is sufficient for soil stabilization considering the compression index. For 2% and 3% LS, the decrease in compression index is approaching constant close to 0.15 at dry and wet side of optimum. As the UCS (Fig. 7b) of the treated soil is significantly higher (>50 kPa) than the untreated soil, the optimum dosage of chemical in terms of UCS is 1%LS. It can be observed for 1% LS in the wet side of the optimum significant increase is 100 kPa, while in dry-side of optimum the increase is 30 kPa, and approx. 50 kPa increase for 2% and 3% LS at dry side of optimum. Therefore, the optimum dosage of LS for soil B can be considered at 2% LS.

For soil B, the optimum amount of chemical is 5% LS with least Cc, 0.22 at dry-side of optimum, Fig. 4(a). For UCS (Fig. 8(a)), the maximum increase is 40 kPa at 1% LS for dry-side of optimum and 30 kPa at wet-side of optimum. From Table 3, CIF has found to decrease almost 50% for 3% LS and 68% for 5% LS. Since, the problem of swelling and shrinkage is the main concern here, the optimum amount of LS can be fixed at 5% LS having least FSI, least swelling (%), permissible Cc, permissible UCS and least shrinkage.

4 Conclusion

The efficacy of using lignosulfonate in stabilizing the swelling soils was assessed in this study. The index and the engineering properties of untreated and treated soils were evaluated. Based on the experimental results, following conclusions are made:

1. Two soils with different plasticity index are used to evaluate the efficacy of lignosulfonate for swelling soils. The plasticity index decreased from 68% (soil A) and 93% (soil B) to 5.74% and 13.2% respectively. This leads to change in soil type from highly plastic (CH) to MI-MH (soil A) at 5.74% PI or MH, soil B.
2. Free-swell index decreased from 90% (high swelling) to 40% (medium-low swelling) at 2% and 3% LS for soil A. For soil B, the FSI decreased from 105% to 50% at 5% LS. While the swell pressure under compaction state, is found to reduce from 161 kPa to 31 kPa at dry side of optimum and 30 kPa to 10 kPa at wet side of optimum for 1% LS and 10 kPa for 2% and 3% LS addition for soil A. While for soil B, swell-pressure decreases from 200 kPa to 40.5 kPa and 110 kPa to 30 kPa at wet-side of optimum. Decrease in DDL results in decrease of swell pressure as a result of decrease in the thickness.
3. From the unconfined compressive test, for soil B the UCS increase is less as compared to soil A. The reason could be due to increase in amount of bentonite content. The clay

content% need to be ascertained upto which lignosulfonate efficiency is significant. It can be observed for soil A, with increasing the LS-percentage from 0 to 3, the UCS increase from 400 kPa to 460 kPa (i.e., the relative maximum increment was up to 15%) at dry-side of optimum and 328 to 565 kPa (72% increase) at wet-side of optimum. Soil B has increase of 30 kPa (1% LS) at dry-side of optimum and 10 kPa at wet-side of optimum. The major contribution to this increase in UCS is attributed to the formation of coat around the particles which prevent water to come in contact with the smectite group in the soil.

4. It can be concluded that for soil A, 2% LS addition to soil changes the soil type from CH to MI decreasing swell-index from 90-40% bringing the soil type to silty. The swell% has been decreased from 9.9% to 4% more than 50% at dry side of optimum and negligible swelling at wet side of optimum. The swell% is observed to follow a constant trend after increase beyond 2% LS. The compression index for 2% LS addition is 50% from 0.473 to 0.127 at dry-side of optimum and wet side of optimum has Cc value of 0.15. UCS has found to be 50 kPa more than untreated at dry-side of optimum and 100 kPa after 28 days of curing. Therefore addition of 2% lignosulfonate to the soil is found to be optimum dosage according to swell %, permeability, compression index and strength enhancement.

Soil B has changed from CH (140%) to MH (50%). Soil Strength has found to increase, but the strength is not significant (>50 kPa) at both dry- and wet-side of optimum. Swell (%) has found to decrease from 15% to 3% (5% LS). Compression index has also found to decrease by 50% at 5% LS. The amount of LS for soil A is found to be optimum with 3% swell (%) at 5% LS at dry-side of optimum, crack intensity factor reduced to half as compared to untreated soil. It is observed for both the soils of different swelling characteristics, lignosulfonate has enhanced the characteristics.

It is imperative for further research to establish the criteria for the usage of lignosulfonate according to clay-content of soils. Here research has been limited to two soils of different clay-content for testing the efficiency of lignosulfonate. Criteria for choosing the chemical content (LS) with addition of additives in small amount like fly-ash can enhance the soil behavior with less harm to the environment.

References

Alazigha, D.P., Indraratna, B., Vinod, J.S., Ezeajugh, L.E.: The swelling behaviour of lignosulfonate-treated expansive soil. Proc. Inst. Civil Eng. Gr. Improv. **169**(3), 182–193 (2016). https://doi.org/10.1680/jgrim.15.00002

Ta'negonbadi, B., Noorzad, R.: Stabilization of clayey soil using lignosulfonate. Transp. Geotech. **12**, 45–55 (2017). https://doi.org/10.1016/j.trgeo.2017.08.004.

Alazigha, D.P., Vinod, J.S., Indraratna, B., Heitor, A.: Potential use of lignosulfonate for expansive soil stabilization. Environ. Geotech. **6**, 480–488 (2017)

Younus, M.M., Sreedeep, S.: Evaluation of bentonite-fly ash mix for its application in landfill liners. J. Test. Eval. **40**(3), 357–362 (2012). https://doi.org/10.1520/JTE104161

Mollamahmutoğlu, M., Yilmaz, Y.: Potential use of fly ash and bentonite mixture as liner or cover at waste disposal areas. Environ. Geol. **40**(11–12), 1316–1324 (2001). https://doi.org/10.1007/s002540100355

BSI BS 1377: Methods of test for soils for civil engineering purposes. Part 2: Classification Tests. British Standards Institution (BSI), London (1990)

ASTM D4318, ASTM D 4318–10, and A. D4318–05: Standard test methods for liquid limit, plastic limit, and plasticity index of soils. Report **4**, 1–14 (2005). https://doi.org/10.1520/D4318-10

ASTM: test method for laboratory compaction characteristics of soils using standard effort (12.400 ft-lbf/ft 3 (600 kN-m/m 3)), ASTM D698–12. ASTM Int. 1–13 (2012). https://doi.org/10.1520/D0698-12E01.1

Pycnometer, W., Purposes, G., Barrel, S., Soil, T., Limit, P., Ovens, V.: Standard test methods for one-dimensional swell or collapse of cohesive soils 1. ASTM Int. **D4546-14**, 1–9 (2011). https://doi.org/10.1520/D4546-08.2

ASTM: One-dimensional consolidation properties of soils using. ASTM Int. **11** 1–10 (2011). https://doi.org/10.1520/D2435

ASTM: ASTM, D 2166–13, standard test method for unconfined compressive strength of cohesive soil, ASTM international, West Conshohocken, PA, 2013. ASTM Int. 1–7, (2013). https://doi.org/10.1520/D2166

Jayanthi, P.N.V., Singh, D.N.: Utilization of sustainable materials for soil stabilization: state-of-the-art. Adv. Civil Eng. Mater. **5**, 46–79 (2016)

Tingle, J.S., Santoni, R.L.: Stabilization of clay soils with nontraditional additives. Transp. Res. Rec. **II**, 72–84 (2003)

Villar, M.V., Lloret, A.: Influence of dry density and water content on the swelling of a compacted bentonite. Appl. Clay Sci. **39**, 38–49 (2008)

ASTM D6913–04R2009: Standard test methods for particle-size distribution (gradation) of soils using sieve analysis. In: ASTM International West Conshohocken, PA, vol. 4, pp. 1–35 (2004)

Bolt, G.H.: Physico-chemical analysis of the compressibility of pure clays. Geotechnique **VI**, 86–93 (1956)

Wroth, C.P., Wood, D.M.: The correlation of index properties with some basic engineering properties of soils. Can. Geotech. J. **15**(2), 137–145 (1978)

Nguyen, L., Fatahi, B., Khabbaz, H.: Development of a constitutive model to predict the behavior of cement-treated clay during cementation degradation: C3 model. Int. J. Geomech. **17**(7), 04017010 (2017)

Alazigha, D.P., Indraratna, B., Vinod, J.S., Heitor, A.: Mechanisms of stabilization of expansive soil with lignosulfonate admixture. Trans. Geotech. **14**, 81–92 (2018). https://doi.org/10.1016/j.trgeo.2017.11.001

Pal, S.K., Ghosh, A.: Volume change behavior of fly ash-montmorillonite clay mixtures. Inter. J. Geomech. **14**(1), 59–68 (2014). https://doi.org/10.1061/(ASCE)GM.1943-5622.0000300

Phanikumar, B.R., Sharma, R.S.: Volume change behavior of fly ash-stabilized clays. J. Mat. Civil Eng. **19**(1), 67–74 (2007). https://doi.org/10.1061/(asce)0899-1561(2007)19:1(67)

Soltani, A., Deng, A., Taheri, A., Mirzababaei, M.: A sulphonated oil for stabilisation of expansive soils. Inter. J. Pavement Eng. **20**(11), 1285–1298 (2019). https://doi.org/10.1080/10298436.2017.1408270

Brandl, H.: Mineral liners for hazardous waste containment. Geotech. **42**(1), 57–65 (1992)

Author Index

B. Fatahi et al. (Eds.): GeoChina 2021, SUCI, pp. 179–180, 2021.
https://doi.org/10.1007/978-3-030-79854-3